HJSR

50th Anniversary Edition

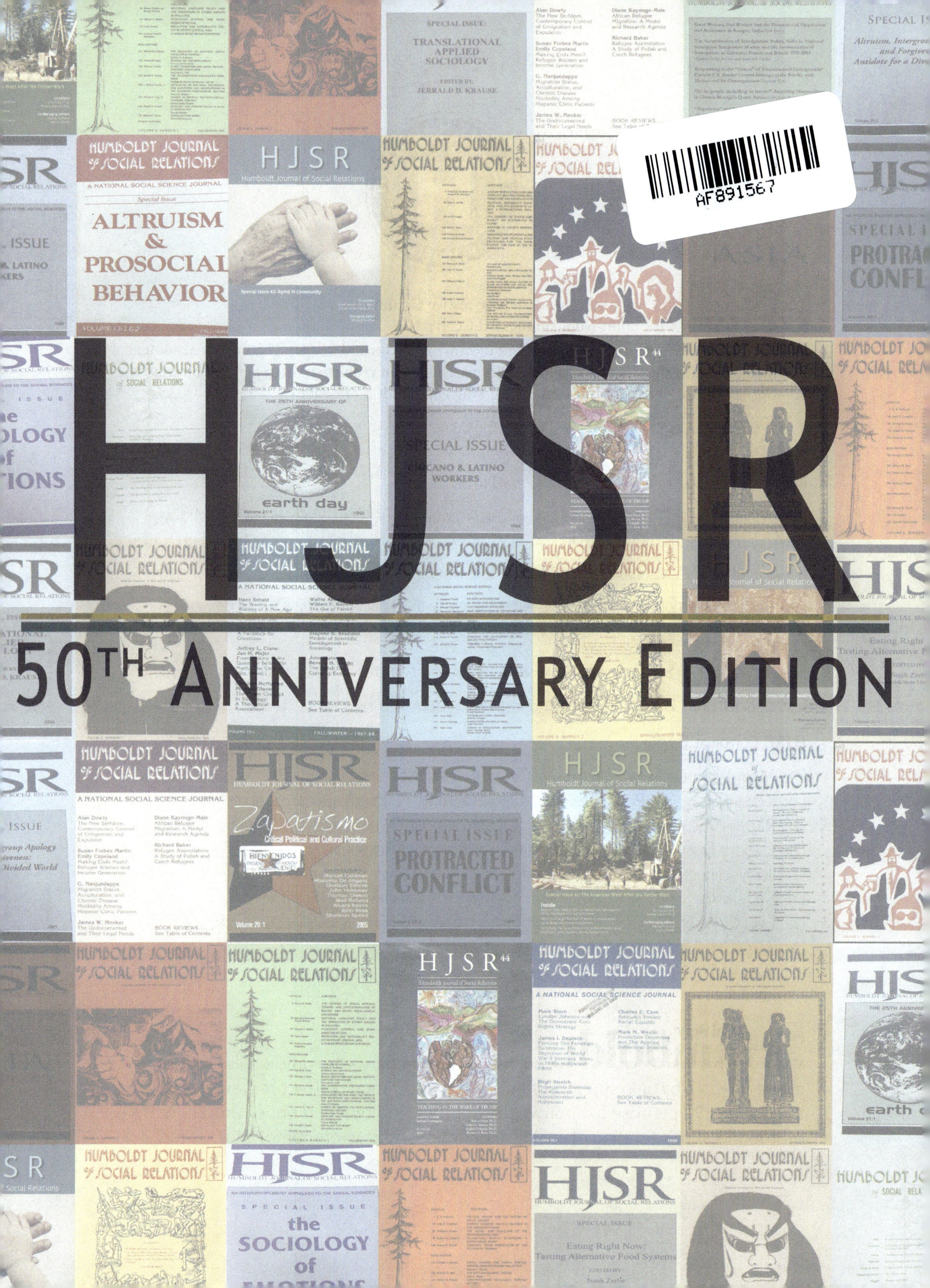

The Press at Cal Poly Humboldt

Cal Poly Humboldt Library
1 Harpst St.
Arcata, CA 95524-8299
press.humboldt.edu
press@humboldt.edu

ISBN: 978-1-947112-90-2

Front Cover Image Proivided by Cal Poly Humboldt

Back Cover Image "Pepperwood Flat on the Avenue of the Giants" from the Tim McKay Postcard Collection Provided by Cal Poly Humboldt University Library Special Collections

Previous *HJSR* Cover Images Provided by Cal Poly Humboldt Department of Sociology

Humboldt County Peacemaker Prize Nomination and Photos of Sam Oliner Graciously

Provided by Steve Stamnes

Special thanks to Sponsored Programs Foundation for sponsoring this edition of *HSJR*

Cover & Interior Layout / Design by Alannah Guevara

The Press at Cal Poly Humboldt. publishes high-quality scholarly, intellectual, and creative works by or in support of our campus community. The press supports the university's purpose to improve the human condition and our environment by promoting understanding of social, economic, and environmental issues.

HJSR

50TH ANNIVERSARY EDITION

Humboldt Journal of Social Relations

50th Anniversary Edition: Becoming a Polytechnic

THE PRESS AT CAL POLY HUMBOLDT

The mural on the cover of HJSR was created in 2015 by Humboldt students working with Saba, a New Mexico-based indigenous artist. The collaboration with Saba arose from a student protest of a painting that showed Mexicans as kitchen laborers coupled with the lack of other images of BIPOC people in art displayed across campus. The cover mural promotes art and culture by and for people of color. The images in the cover mural itself were/are sites of controversy and interpretative tension. The editors chose the mural for the cover because of the visual links to many of the themes of the articles in this issue, as well as the role that students played in its creation. While this issue includes interviews with Art + Film faculty, this mural was not a project of the Art Department. The artwork is located prominently near the entrance to the Cal Poly Humboldt Gutswurrak Student Activities Center. To learn more about the mural and the context surrounding its creation, visit the following links:

- *El Leñador news article:* https://digitalcommons.humboldt.edu/cgi/viewcontent.cgi?article=1006&context=el-lenadornews2015

- *Humboldt Now press release:* https://now.humboldt.edu/news/university-center-dresses-up-with-student-led-mural

- *Protest document with demands for mural, better representation in University admin, and better wages for employed students of color:* https://library.humboldt.edu/art/Artists/Spaulding_Ryan/No%20need%20for%20Super%20Taco%20unless%20its.pdf

In memory of
Samuel P. Oliner

1930-2021

Table of Contents

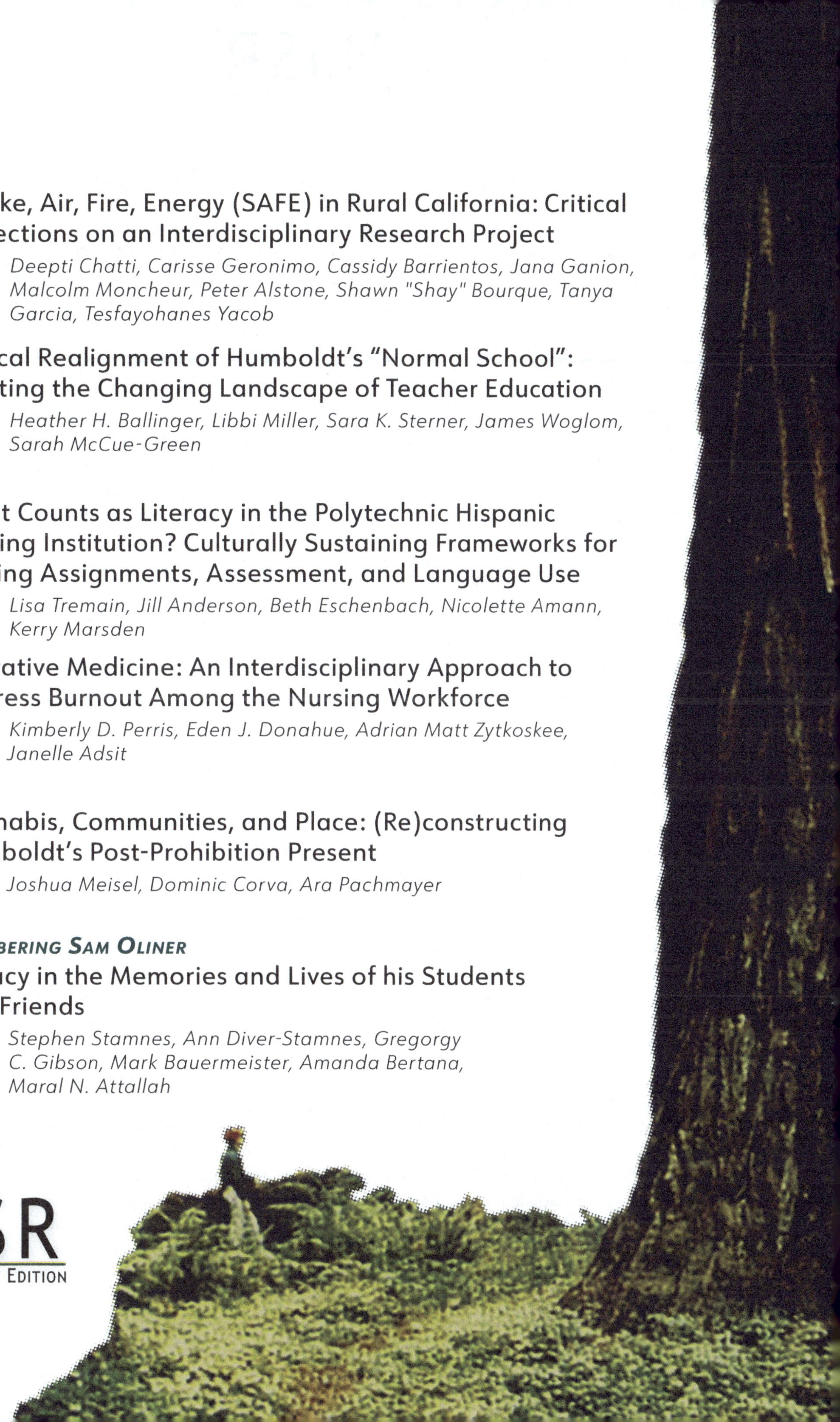

HJSR
50TH ANNIVERSARY EDITION

Introduction from the Editors

Mary Virnoche <mary.virnoche@humboldt.edu> *(Cal Poly Humboldt)*, **Jennifer Eichstedt** *(Cal Poly Humboldt)*, **Erin C. Kelly** *(Cal Poly Humboldt)*, **Kyle Morgan** *(Cal Poly Humboldt)*

This 50th Anniversary Edition of the *Humboldt Journal of Social Relations* (HJSR) was conceived in both challenging and exciting times. Amidst the second year of the COVID-19 pandemic, as well as the launch of Humboldt as the third polytechnic in the California State University system, Editorial Board members discussed leveraging HJSR to both highlight and catalyze exciting local scholarship.

As editors of this issue, we are thrilled with the outcomes. We congratulate the faculty, staff, students, and community partners who took up our challenge to:

> construct snapshots of the Humboldt intellectual landscape at a pivotal moment in the university history, as the campus transitions to a polytechnic institution with a core identity as an HSI and MSI. Co-authored articles across subfields will position their narratives at the nexus of changes in the field, integrating threads of past Humboldt faculty scholarship, and building forward thinking narratives about current research, applied partnerships, and pivotal questions shaping the work to be done.

This issue intentionally seeded multidisciplinary conversations that wrestle with pressing local and regional problems. The work sets the stage for creative, forward-thinking scholarly collaborations that are the signature of Cal Poly Humboldt. Some of the author teams include partners from outside the university, including Tribes, agencies, and non-profit organizations. Across the issue, authors also detail connections between research and the university commitment to hands-on education that is a cornerstone of the Humboldt student experience.

In the opening essay, Swartz narrates a retrospective on HJSR founder and sociology faculty member Samuel Oliner. As a Holocaust survivor, Oliner dedicated his career to the study of altruism, kindness, apology, and forgiveness, often in collaboration with his lifelong partner Pearl Oliner. As he passed on leadership of the Altruistic Behavior Institute to Swartz, Oliner was committed to translating that work into action, addressing a broad range of challenges from homelessness and drug addiction to nuclear disarmament and COVID-19. Oliner's work was foundational to the department commitment to applied social science scholarship that now aligns with the new university status as a polytechnic.

In the next essay, HJSR managing editor Guevara invites us into conversations with Department of Art+Film faculty members. Weaving narratives from a series of interviews into the essay, she details the work of faculty and students as they embrace the polytechnic identity. From studio arts to museum and gallery practices, hands-on learning has been and will continue to be central to the everyday lives of art faculty and students. Photographs captured by students accompany the essay and provide examples of these hands-on experiences. The subject matter of the photographs further illustrates the connections between department faculty and students and the polytechnic work across programs.

Native American Studies lays at the heart of a movement to transition Humboldt to a polytechnic university. In their article *Polytech to PolyTEK*, Risling Baldy, Reed, and Begay narrate many threads of Native American Studies (NAS) woven across the polytechnic transformation. The article conceptualizes Traditional Ecological Knowledge (TEK) and details projects and programs across the university in which NAS curriculum and leadership is central. Many academic majors ground their programs in the physical and cultural places we call Humboldt. The authors detail past and current relationships with indigenous peoples of this land, and the significance of indigenous ways of knowing to solve local and regional challenges.

In *Wildfire*, Kane et al. explore partnerships developed to explore and combat wildfires in the northern California region. They describe several projects that seek to build resilience to wildfire by incorporating multiple ways of understanding and responding. Through founding the Fire Resilience Institute, as well as the development of a Bachelor of Science in Applied Fire Science and Management, the work addresses "the grand wildfire-related challenges" of the present and future. In particular, the authors focus on work to increase a fire resilience workforce, interconnect fire knowledge systems, and increase the region's resilience to wildfire.

Taking up another facet of climate change in *Sea-Level, Rise* Richmond et al. describe a partnership between the Wiyot Tribe, government agencies, and people from across departments at Cal Poly Humboldt to addresses this challenge in Wigi (Humboldt Bay). Wigi is part of the homeland of Wiyot people, whose descendants are members of the Wiyot Tribe, as well as the Blue Lake, Bear River, and Trinidad Rancherias. The number of authors on this article reflect this remarkable partnership, with 25 contributors from across multiple disciplines and affiliations. The authors outline the risks of sea-level rise to Wigi, an area marked by colonization, displacement, and development. To address the risks of sea-level rise, they narrate the importance of indigenous knowledge, alongside geospatial analysis, engineering, and social sciences.

Next, in *Smoke, Air, Fire, Energy (SAFE)*, Chatti et al. describe an ongoing partnership between Cal Poly Humboldt, the Schatz Energy Research Center, the Karuk Tribe, and Blue Lake Rancheria for community-engaged research at the intersection of technology, fire, and indigenous knowledge. The partnership works to honor Tribal sovereignty and expertise, while incorporating engineering and social science knowledge across several research projects designed to accelerate the adoption of climate-resilient technology, smoke detection, and energy microgrids.

Focusing our attention on *Teacher Education*, Ballinger et al. reconceptualize and reaffirm a vision for teacher education that is reflective, relational, reconstructive, and research-oriented. These commitments integrate concepts of positionality, communities of practice, community partners, and polytechnic perspectives to address issues of inequity and oppression, deprofessionalization, and teacher shortages. The vision supports teacher and student alike, and holds the promise of innovation and sustainability in university coursework and clinical practice.

Continuing a focus on education, Tremain et al. in *What Counts as Literacy?* challenge the reader on fundamental assumptions of how they define writing, how they learned to write in their discipline, and how those assumptions can be intentionally addressed to create more inclusive and culturally sustaining disciplinary literacy development. They theorize and describe classroom pedagogies that support the development of student disciplinary literacies through culturally sustaining and socially just approaches. The authors also suggest strategies for writing assignment and assessment designs that support student multilingual and multiliterate ways of knowing.

The COVID-19 pandemic amplified health care provider (HCP) and related system challenges for addressing HCP emotional and social psychological strain. In *Narrative Medicine*, Perris et. al. detail the impact of this trauma on HCPs, comparing it to post-traumatic stress syndrome of military personnel returning from combat. In the article, Cal Poly Humboldt nursing faculty, in collaboration with English faculty, provide a theoretical and evidence-based case for changes in nurse training programs to mitigate HCP trauma and infuse narrative medicine tools into the nursing workforce. They report in detail on narrative medicine as practiced in the Cal Poly Humboldt nursing curriculum. This work has broader implications for curricular (re)design of current and future programs in the School of Applied Health and the scholarship of teaching and learning in the field.

In *Cannabis, Communities, and Place*, Meisel, Corva, and Pachmayer describe the changing role of cannabis both as a topic of scholarly interest at the university, and as an economic driver in Humboldt County. The authors highlight the new Cannabis Studies Program at Cal Poly Humboldt, which takes an interdisciplinary approach to the complex cultural, social, and ecological impacts of cannabis production and cannabis policies. They follow this with an outline of the history of cannabis in the region, ending with the post-prohibition era "crossroads" for the industry. While cannabis cultivation has become widespread across the state, growers in Humboldt County have suffered as a result of declining prices and a difficult regulatory environment, with consequences. These experiences have consequences for both the county and the university.

On the final pages of this issue, we return again to HJSR founder Samuel Oliner, as witnessed by generations of students. In these closing testimonials, alumni—who are now faculty across the country—share their memories of Oliner's impact on their lives. These alumni, and thousands of others, are a part of the remarkable legacy that has been and will continue to be Humboldt.

HJSR

50TH ANNIVERSARY EDITION

Reflections on Sam Oliner and the Altruistic Behavior Institute

Ronnie Swartz <ronnie.swartz@humboldt.edu> *(Altruistic Behavior Institute; Cal Poly Humboldt)*

THIS YEAR MARKS THE 50TH ANNIVERSARY of the *Humboldt Journal of Social Relations*, founded by Dr. Samuel P. Oliner. Sam joined the Sociology faculty in 1971. While he retired formally from his position in 2000, he remained an active Professor Emeritus through 2010, mentoring new students and faculty and involving many with his never ending commitment to the study of altruism.

As a Holocaust survivor, Sam chose to focus on understanding people like those who had helped him escape. To that end, with his wife Dr. Pearl Oliner, in 1982 he founded The Altruistic Personality and Prosocial Behavior Institute. The Institute celebrated its 40th Anniversary in 2022. Now known as the Altruistic Behavior Institute, it began as the home for Sam and Pearl's pioneering research on characteristics of people who rescued Jews and other persecuted people during the Holocaust. They published their findings in The Altruistic Personality: Rescuers of Jews in Nazi Europe (1988). Social science related to altruism saw a resurgence soon thereafter.

Sam, Pearl, and the Institute went on to explore phenomena that continue to shape everyday life, including forgiveness and apology, love and compassion, and heroism and moral exemplars. That research informed the books *Do Unto Others: Extraordinary Acts of Ordinary People* (2003) and *The Nature of Good and Evil* (2011). Sam's account of his formative years and the lived experiences that brought him to the field of study to which he dedicated his life can be found in academic and popular publications, such as *Restless Memories: Recollections of the Holocaust Years* (1979). Sam's life and work have even been featured in a children's magazine (Lynn 2018).

In 2021, I spent time with Sam Oliner about a month before his passing. He was eager to find ways to get the word out about his latest book, *What Kind of Future Will Our Children Inherit?* (2021). On the back cover of this edited volume that had been a focus for a decade, he wrote:

> [My previous books] concerned altruism, kindness, empathy, and moral responsibility for diverse others. In this book we concentrate on the areas of greatest concern regarding our future as a species... My own work indicates that goodness, defined as concern for others and for making the world a better place, is on the rise.

Sam's commitment to research and action did not waiver across a professional career that lasted over six decades. By the end of our meal, Sam was talking about his next book.

Sam and Pearl moved away from the physical home of the *Humboldt Journal of Social Relations* and the Altruistic Behavior Institute in 2010. That year, Sam asked me to lead the institute in a new direction he envisioned. He wanted it to shift from altruism-related research into altruism-related projects. Now, it is not like he saw these as discrete areas: see, for example, *Toward a Caring Society: Ideas Into Action* (1995). But he wanted the Institute to be more involved in practice. This started with a community-driven effort focused on serving people dealing with homelessness. The work has gone on to include projects aiming to raise awareness about the persistent threat of nuclear weapons, improving outcomes for people who engage in panhandling, training for providers on opioid overdose prevention and intervention, and a podcast about parents caring for young children during the early months of the COVID-19 public health emergency.

This year, the Altruistic Behavior Institute continued a four-year federal grant to strengthen integration of primary health care and behavioral health care. Thanks to support from the Cal Poly Humboldt Sponsored Programs Foundation, we launched an Oliner Altruism Research Archive. This major initiative will make Sam and Pearl Oliner's original research, including audio recordings with Holocaust rescuers and survivors, freely available to researchers, practitioners, and broader communities around the world.

I came to know Sam in the last decades of his life. His warmth toward me, identification of my skills, clarity that there are specific things I should be doing with my time, and confidence that I could be successful continue to shape my personal and professional life. Sam was a strong supporter of the projects I have initiated on behalf of the institute. He saw the ideas he promulgated as being operationalized through the recent work of the Altruistic Behavior Institute.

References

Lynn, Diana. 2018. "The Measure of a Hero: Why Do Some People Risk their own Lives to Save Others?" *Muse* May/June:10-15.

Oliner, Pearl, and Samuel P. Oliner. 1995. *Toward a Caring Society: Ideas Into Action*. London, UK: Praeger.

Oliner, Samuel P. 1979. *Restless Memories: Recollections of the Holocaust Years*. Berkeley, CA: Judah L. Magnes Museum.

Oliner, Samuel P. 2003. *Do Unto Others: Extraordinary Acts of Ordinary People*. Boulder, CO: Westview Press.

Oliner, Samuel P. 2011. *The Nature of Good and Evil*. St. Paul, MN: Paragon House.

Oliner, Samuel P., and Pearl Oliner. 1988. *The Altruistic Personality: Rescuers of Jews in Nazi Europe*. New York: The Free Press

Oliner, Samuel P., and Ronnie Swartz, eds. 2021. *What Kind of Future Will Our Children Inherit?* Arcata, CA: Humboldt State University Press.

Sam and Pearl's groundbreaking 1988 book that led the resurgence of altruistism in the social sciences.

Edited in association with Ronnie Swartz, Sam's final book explores the legacy of his field and the work yet to be done.

Conversations with Art Faculty: Visualizing a Polytechnic "Aesthetic Toolbox" for Hands-On, Community-Based Education

Alannah Guevara *in conversation with* ***Brandice Guerra, Berit Potter,*** *and* ***James Woglom; Julie Alderson*** *in conversation with* ***Nicole Jean Hill***

In early 2023, I had the distinct honor of speaking with three Cal Poly Humboldt faculty members from the Department of Art + Film in the College of Arts, Humanities & Social Sciences. We discussed the theoretical, pedagogical, and interdisciplinary position(s) of art within the vast and deep framework that is "polytechnic." We also considered the flourishing possibilities of growth that accompany the Cal Poly designation. Faculty voices in the narratives include Associate Professor Brandice Guerra, Assistant Professor Berit Potter, and School of Education Interim Department Chair and Associate Professor James Woglom. Though our discussions were tethered to a prewritten interview guide, each interview took on its own unique tone and highlighted a range of challenges and opportunities. When placed in conversation with each other, these interviews provided an exciting vision of the arts at our polytechnic university.

What is a Polytechnic, Anyway?

The title of this special issue is Becoming a Polytechnic. Throughout these interviews, one recurring idea became clear to me: "Becoming" is not a process that occurs overnight. It is ongoing. It both preceded and continues past the official designation of California Polytechnic University, which did occur—seemingly overnight—in February 2022. As French philosophers Deleuze and Guattari posit, "becoming has neither beginning nor end, departure nor arrival, origin nor destination" (1987:342). And so, we might consider our university's "becoming" to be a transition that has always been in process. Indeed, within this issue you will find evidence of this. From the completed work that is highlighted within these pages to the work that is being accomplished at this very moment, it is apparent why Humboldt was selected to become polytechnic.

Another theme across the interviews was that, early in the process of becoming a polytechnic, there was a wide range of anticipatory emotions, from excitement to anxiety—likely throughout all the arts and humanities—regarding the polytechnic designation. It boiled down to one question: What place would the arts have within a STEM-focused university?

Our university was founded in 1913 as Humboldt State Normal School for the purposes of "training and education of teachers and others in the art of instructing and governing the public schools of this state" (Johnson 1913 cited in Tanner 1993:3). Our history is one rich with its focus on technical career training. We approach hands-on vocational training and career preparation as fundamental aspects of higher education. The Cal State University system, which we belong to, provides the following definition for polytechnic: "A focus on applied learning is what sets a polytechnic university apart from a traditional university. Also known as experiential learning, it combines the in-depth study found at universities with practical, technology-based skills training" (2022). This combination of vocational training and experiential-learning is precisely what conjoins the worlds of science and art. As Provost and Vice President for Academic Affairs Dr. Jenn Capps assured:

> We have really strong arts [and] humanities programs, social sciences programs and music programs [...] We are incorporating an interdisciplinary sort of approach to STEM education, where faculty are reaching across disciplines and figuring out how to bring elements of all of these things into academic programs. We are being open and recognizing that critical thinking, communication and the arts make a better engineer, scientist—you name it. (cited in Cal State University 2022).

This is also the reason that the arts will flourish in what many mistakenly believe to be a STEM-centered institution. We are—and have been—preparing our artists with the necessary skills that they will need to succeed in their careers. And we put artists and poets in conversation with scientists and engineers to solve complex problems facing the world.

Let's Begin with an Introduction to Our Art Faculty Interviewees

Brandice Guerra finds beauty and interest within the minutiae of the natural world. Principally a painter, she enjoys creating at the small-scale, and employing traditional, indirect painting methods to achieve her vision. Unable to recall a time when she wasn't drawing, Guerra decided to pursue a career in art as a young adult. She believes that many people who pursue the arts, like her, are wired for it. In 2013, when she first came to, what was then still Humboldt State University (HSU), she was impressed with the competency, skill, and beauty that she found in the work being produced by the students. Guerra found the Art + Film Department's pedagogical methods to be somewhat unique due to its focus on traditional approaches to art and hands-on teaching. Methods that she states are not present at every university. And then, of course, there's the trees. Cal Poly Humboldt is nestled up against a Northern California redwood forest—the perfect backdrop for any artist wanting to be enveloped by the natural environment. She also notes the seclusion that comes with Northern California living, which is something she believes lends itself to fostering the unique traditions found in the Art + Film Department. As Guerra explains:

> Remoteness can be a very good thing a lot of the time, I think, for our programs. They can be their own thing and have their own kind of approach to things without falling in lockstep with what might be going on in larger cities. I think it's important to have a diversity of approaches and spaces for students to choose from when they're kind of picking a school.

Berit Potter studied art history and psychology at UC Santa Cruz, where she learned about art worlds elsewhere, particularly in New York. After completing several internships in the Bay Area, she promptly relocated to earn her master's in museum studies at New York University and found employment with the Whitney Museum of American Art. After deciding to pursue a career in museum work, she earned her PhD in art history. By the time she finished her degree, she was considering becoming a teacher instead, and eventually, she settled on doing both. In an ironic twist, it was her time and studies in New York that finally taught her about the art happening in California, which she returned to in 2013. She spent the next few years teaching at several different colleges until 2018 when began working at HSU. She counts herself lucky to have found her position in the Art + Film Department, which affords her the opportunity to share her passion for museums and guide students towards their chosen careers. Reflecting on her own time as a student, Potter explains:

> That's something that I didn't really get in my time as an undergrad. I had ideas about what I was interested in, but I didn't really know how to pursue them. And I hope that through the museum and gallery practices certificate, the students get a better sense of not only what they want to pursue, but how they can go about making that happen.

James Woglom, a self-proclaimed artist from birth, first found his love of creating art through crayons. Encouraged by his teachers after drawing Darth Vader, he held his first art show in early elementary school. After that, it was off to the races. Woglom pursued his passion at university, beginning with a bachelor's in studio arts with a focus on drawing and painting, and then earning a master's and a PhD in art education. In 2015, Woglom joined the Art + Film Department at HSU. He was attracted to it because:

> I liked the focus on issues of social justice and the mission of the university in its learning objectives, and honestly, I had just never seen that sort of overt commitment to two specific issues from a state university. I don't know, this might just be a presumption, but I had seen more generalist job calls coming from state universities, and this was, I felt like, people doing the work. So exciting.

Guevara: *Tell me a bit about your position and what you do in the Cal Poly Humboldt Art Department.*

Woglom: I am the Professor in Arts Education. Our arts certification program is an undergraduate teacher pathway. These are—what would I call it? It's not quite a credential, but it's an authorization put forth by the CTC, which is the California Commission for Teacher Credentialing. It prepares students and curriculum in subject matter so that they can then go on to a credential program.

So, whereas studio majors would focus on probably a couple of different media that they were super interested in before diving into specifics, or art history students would focus on how to make art then a bunch of looking at art history, in art education, we cover all of the media that could potentially be taught in K-12 schools and colleges in California. We then look at art history as a kind of breadth thing—making sure that students have a widely informed knowledge of what the arts are so they can take that to as many students as they possibly can over their teaching careers.

Guerra: As somebody who teaches in a kind of a humanities-oriented, liberal arts school, I think my primary job is to expose people to images and ideas. I hope through that activity they can find access to beauty, things that bring them happiness, and to ideas that help them navigate their lives as members of communities. And that's kind of like the big overarching thing.

And then, in order to do that, I'm sharing images, I'm sharing ideas. I'm also teaching people how to use tools and refine their own motor skills and skills of observation; to use their own bodies and the tools that I'm teaching them how to use to communicate their own ideas. I encourage them to kind of refine their own skills so that they can then produce their own images and share their own ideas. Because in the end, that's kind of what a lot of art making is about. It's sharing ideas, whether it's embedded in the object itself or whether it needs external explanation.

What we're doing in the Art + Film Department is integrally engaged with the community in a way that I think is very much in keeping with the polytechnic model.

Cal Poly Humboldt Art Education students, along with students from Redwood Coast Montessori High School and local community members, paint a 20'x250' mural on the northern exterior wall of the high school's building at the corner of 8th and K St. in Arcata. Photo credit: Alisa Ibyadullin

Potter: I teach art history and museum studies classes in the Art + Film Department. I also oversee the certificate in museum and gallery practices, which is interdisciplinary. It has tracks for art, anthropology, history, environmental science, management—which is our newest one—and Native American studies. I'm also the faculty coordinator for our new place-based learning community (PBLC) called the Creative Coast. I get to do the practice-based work of helping students install shows and then also talk about the history and theory of museums. I really love it.

My research focuses on the history of U.S. museums in the 20th century, as well as the development of modern art in relation to San Francisco and the West Coast more broadly. I also think about cultural exchange between U.S. and Latin American artists and art institutions in the 20th century. I try to fold that research into my classes as much as possible. So, I teach classes around Latin American modernism, museums and the politics of display, museums and gallery practices, and so on.

Guevara: *In November 2020 the Chancellor invited Humboldt to prepare a prospectus for becoming a polytechnic. Can you recall your initial reactions to that announcement? Among the art faculty, staff, and students—what were some of those early conversations and questions?*

Woglom: Honestly, when I read it, I had a vague understanding of what a polytechnic school was. I knew it was more STEM, or at least had a sense that it was more of a STEM focused endeavor. But when I looked at the definition and looked at a bunch of descriptions of polytechnic universities, I found that one of the more central ideas is that it's an experiential-based learning model. Hands-on learning is at the core of the curriculum. And I think that, honestly, that's what we do in a lot of the Art + Film Department. Not just in the media areas but in our museum and gallery practices, or going out and working with the community and galleries and museums to create more opportunities for showing and talking about art.

Then, from talking to folks, I think there was an overwhelmingly positive feeling around the polytechnic shift and what it could do with, you know, concerns about how it might play out on, say, a disciplinary level, among all sorts of possible and potentially scary change. In art education, our entire program is sort of service-learning centered. It's based around this idea of learning to become teachers by going into spaces where there is a dearth of arts-based programing, writing programing for those spaces, and then teaching it or helping to teach it. So yeah, I think that hands on stuff is already so much a part of art education as a field. Specifically, what we're doing in the Art + Film Department is integrally engaged with the community in a way that I think is very much in keeping with the polytechnic model.

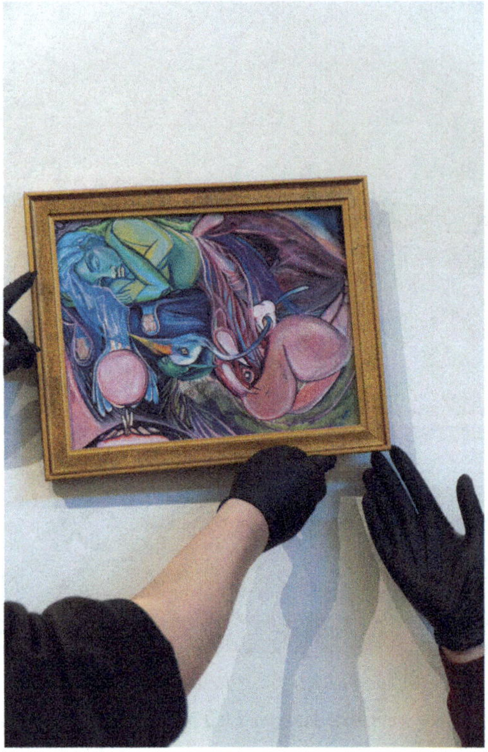

Art 356: Museum and Gallery Practices students install the 2023 Art Graduates Exhibition at the Reese Bullen Gallery. Photo credit: Sonia Sanchez

So, once we talked about what it meant, it started to feel very comfortable. We were already doing a lot of what it meant. I think that's the feeling from a lot of folks in the university; probably like this is already sort of what our approach is. It's sort of codifying that and making it clear to potential future students that that's what's going on here.

Guerra: I think there was maybe a little bit of anxiety within the arts and humanities faculty. I kind of knew that some of the other Polytechs had Art + Film Departments. I wasn't sure that the day-to-day activities of my own job, my own labor, would change. I teach kind of hands-on and skill-based things, which I think fits into technical education pretty readily. I think the concern was that there would be, you know, an increased focus on STEM, and then a kind of scarcity in terms of financial resources for the arts and humanities. Which I do think is kind of a valid concern. But I think things have changed now.

As far as my students go, I think the name change to Cal Poly was kind of a concern. But I don't know if people understood that there would be a shift in priorities.

Potter: I think that a lot of our conversations were centered around the fact that Humboldt becoming a Cal Poly made a lot of sense, given the University's investment in sciences and just how much the school already prioritized hands-on learning. That's something that we obviously do in the Art + Film Department. It's essential for our classes, whether that be jewelry making or photography. Also, almost half of students with a minor in art (25 of 54) come from STEM majors. So, there's already a cross-pollination happening between those areas, which I think is really exciting. It's something that I haven't really seen at the other schools where I've worked. You could think about the scientific drawing classes that are offered in the Department that make use of the Wildlife Museum's collection as examples of that. In so many different ways, there's a lot of collaboration and interdisciplinary practice.

Also, polytechnics tend to have a vocational focus, which is important. I oversee a professional certificate in museum and gallery practices which is geared toward preparing students for work in museums and galleries, and other aspects of the art world. Like I said, it's an interdisciplinary certificate, so, students can work in history museums and archives and other institutions of that nature. I found that really exciting, too. Our certificate involves everything from hands-on work, like installing exhibitions to working with the collection, as well as tasks like writing condition reports for objects. And then the students do internships in the community, which helps them build their careers.

Guevara: *Now that the University has been officially designated a Polytechnic, how have those conversations changed, if at all, since that time? What has become of those hopes and possibilities?*

Guerra: I think that the increased financial support from the state—and I'm not going to speak for everybody because my opinions are my own—has kind of alleviated some of the initial anxiety. People feel a little bit better and safer in their jobs.

I've found that there's an increasing kind of outreach from the sciences. There's a desire for collaboration, which I like. I mean, I've done scientific illustration. I teach a couple of classes in scientific drawing. I've always had an interest in the biological sciences, so I'm kind of excited to collaborate with science people. Right now, I'm working with a colleague in biology to bring a scientific illustration intern to the campus. They'll work with Dr. Peloso, who is the person I'm working with to illustrate recently extinct Brazilian frogs. And the hope is that I could mentor them and provide tools for them, and maybe a space within my classroom for them to work alongside my own students.

We're also excited that there's going to maybe be more economic development in the local area. And maybe more opportunities for local people to get degrees that they might have wanted to get without going far away to get them. And just the kind of economic development in terms of, like, people who are getting engineering degrees or getting computer science degrees and then starting businesses locally. I think that would be wonderful.

Woglom: I think that we're an advantageous school to be taking on this process of thinking through the ideas of the polytech. What's super exciting about this space is this incredible engagement with this beautiful environment. There's also this engagement with community through that social justice curricular lens. I think (we) could potentially become a model for other polytech-

Wildlife Management and Conservation major Sidney Wells sketches from bird specimens in the Wildlife Building. Photo credit: Sonia Sanchez

nics. We could show them that this is another way to look at hands-on, community-based, experiential-centered education.

I'm in the School of Education, as well, and a major focus of ours is thinking about how we can work to create an ongoing kind of a sustainable ecology of teacher preparation. You know, making sure that kids are constantly getting robust, educated experiences. If there is one concern I have, it's making sure that there's an equitable dispersal of arts programing and educational initiatives. It was pretty all over the place for a long time.

While we're on the subject, Prop. 28 passed the vote in November, which requires K-12 public schools to spend at least a certain percentage of their funds in arts and music programing.[1] It won overwhelmingly. But what it effectively does is place ongoing funds into schools requiring art teachers. There's a long history for why California kind of lags behind other states in providing free arts education at that K-12 level. But, I mean, this is a night-and-day shift. Any district in the area that has over 500 students has to hire an art teacher, which is huge. It's speaking to a new employment push that is like, I don't know, it's one of the things I always dreamt might happen, but never imagined it would. And now that it has, it's like, oh, we have to figure out how to prepare and hire a whole bunch of art teachers in the next little bit of time. It's super exciting. It's almost like a perfect storm of interesting changes that I think are going to lead to a lot more people having a lot more access to artful ideas. And hopefully it leads to them making the arts a part of their daily lives.

Art+Film student Cita Hunter sketches in the Cal Poly Humboldt Greenhouse. Photo credit: Alisa Ibyadullin

Potter: Last year, our Department Chair Nicole Jean Hill and I worked on a proposal for the new PBLC, the Creative Coast, which just launched this academic year. Some of the first-year art majors were involved in it this year. By next year, all of them will be involved, including film (students), which is really exciting. It's basically

1. According to Austin Beutner, former Los Angeles Unified Superintendent, "Barely 1 in 5 California public schools has a full-time arts or music teacher [...] Prop. 28 will create more than 15,000 additional jobs for teachers and teachers' aides as well as in community arts organizations. This will help prepare California school children for good-paying jobs, not just in the arts but in other sectors where the creative-thinking and problem-solving skills they learn can be applied [...] Longer term, Prop. 28 will lead to greater diversity in the technology, media and entertainment industries as a broader population of students in California public schools find the doors of opportunity open for them with their newfound skills and experiences" (cited in Jones 2022).

geared towards introducing first-year art students to the University and the Greater Humboldt community through experiential learning. We took a trip to Redwood National Park at the start of the fall semester for part of Summer Immersion,[a] which was really fun. We host artist talks with local muralists like Alme Allen, which is fantastic. We actually got to go see him while he was in the process of creating one of his murals.[b]

Guevara: *As an art faculty and department, what is the collective vision for growth and development that has roots in the polytechnic conversations?*

Potter: I mentioned our new PBLC, the Creative Coast. One of the most exciting things about it is seeing how involved the students are getting in the Department right from the beginning. I am also the faculty advisor for the Student Access Gallery, and typically students don't participate in the exhibitions during their first year. Maybe they're a little bit nervous to participate right from the start. But this year in two of our earliest shows, we had students from the first-year cohort already exhibiting their work in the gallery, and that was really great to see. I think that the Creative Coast is making an impact in terms of helping the students feel more a part of the community from the beginning of their time at Cal Poly Humboldt.

And then we have Assistant Professor Stephen Nachtigall and Nicole Jean Hill working on the creation of the new digital media major, which is really exciting. There's the BFA (Bachelor of Fine Arts), which is fairly new to our department. It focuses on preparing students to become professional artists. In addition to that, the art history concentration is now dedicated to both art history and museum studies, which means it's more actively preparing students for jobs in the field.

Images can communicate complex ideas in a way that is democratic and, you know, allows a wider audience to have access to those ideas.

Guerra: I think, as a department, we deal in both ideas and the hands-on technical preparation that is designed to aid in communicating those ideas. That makes us well-positioned to participate in collaborative initiatives. I don't think that art just exists to provide illustrations for textbooks and things like that, because I think that can be a direction that people tend to think that we want to go in. But it can serve that purpose. Images can communicate complex ideas in a way that is democratic and, you know, allows a wider audience to have access to those ideas.

Then, you know, something might just exist in a journal article where only a very rarefied audience is going to have access to read. And I think with things that are going on in the environment now, it's increasingly important to help people be aware of things. That's where the arts can come in and be useful.

I'm thinking of the many kinds of ShakeAlert notices that I've gotten recently and the friendly design of that app, you know. If you're in a panic mode and trying to read information, then having access to that kind of complex scientific information in a way that's easy to navigate and looks nice can be so important. Somebody had to design that thing, you know. Somebody had to draw the maps and make graphics and sounds and all of that.[c] That's a very clear way that we can be collaborating.

Woglom: Personally, I'd like to look towards including dance and theater in future iterations of arts pathway programming. There are actually new credential authorizations for theater and dance that were just made available in the last couple of years for the first time since 1972. It was something that I was interested in doing, but now I feel like it's imperative. I also would love to expand the interrelationship between the arts and the community and different cur-

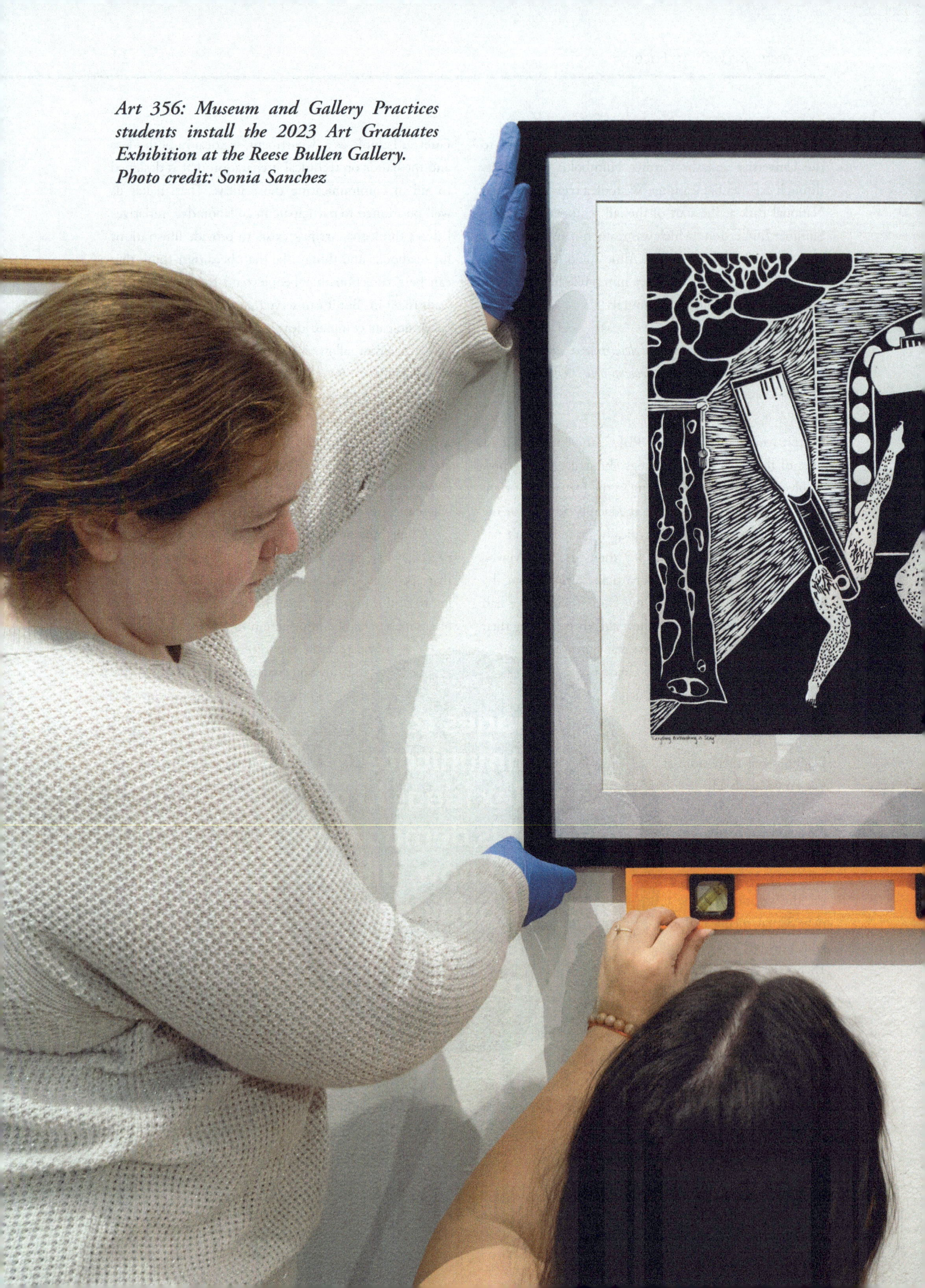

Art 356: Museum and Gallery Practices students install the 2023 Art Graduates Exhibition at the Reese Bullen Gallery. Photo credit: Sonia Sanchez

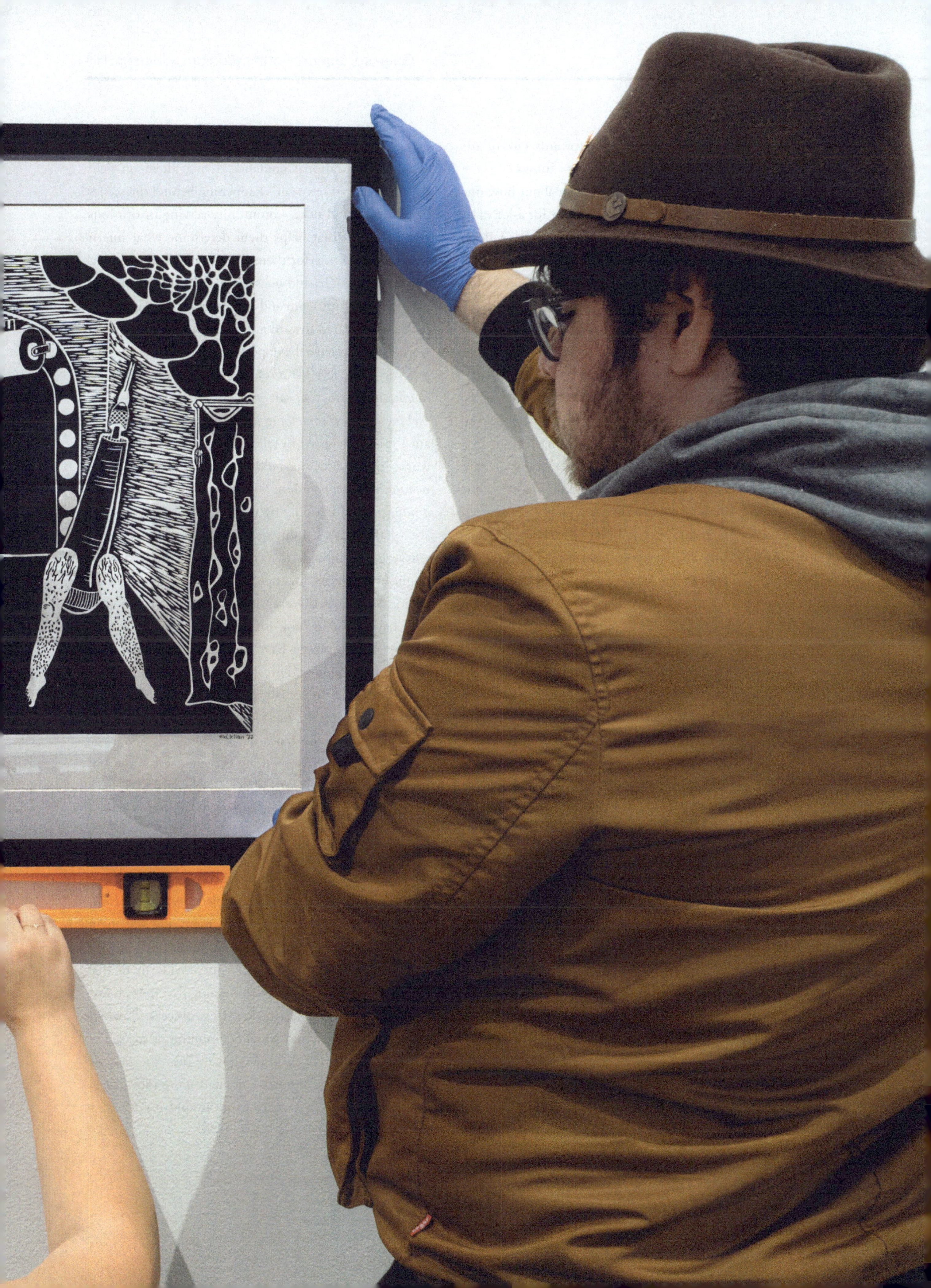

ricula. I mean, we've been working towards a lot of arts integration ideas across a lot of different things.

I was talking to someone recently about how one, I guess, centerpiece of the Venn diagram for a lot of us in Humboldt is aesthetic appreciation, though I don't necessarily think it's overtly spoken to in a lot of our coursework. How can we make aesthetic appreciation of content one more part of the toolbox for anybody engaging with the professional practices or business that they're already doing in any of the sciences? How can we make drawing, painting, photography and the aesthetic appreciation of the world that scientists are engaging with a part of their understanding of those spaces? So, expanding, I think, is a concern within that.

I'm thinking about a lot of the interactions we have in the classroom, as being a kind of relational aesthetic endeavor. Thinking about classroom practice as conceptual art—that's not necessarily my idea, it's something that's been talked about forever. But it's definitely a concern and interest to me. How can I both think about and frame what we're doing together as a classroom community, or programmatic community, as artful action?

Guevara: *Now that we've begun to establish what the Art + Film Department's role within a polytechnic framework is and can be, can you talk a little more about your own ideas about and role(s) within the art world? How do you understand your work in relation to a polytechnic model of education?*

Potter: When I graduated from college, I wasn't prepared to decide what kind of work I wanted to do or how I would go about getting a job through experience or education. I didn't have access to a program like the museum certificate that encourages hands-on work and addresses career pathways. It was difficult for me to decide, you know, if I wanted to do curatorial work or registration or conservation or education or one of the many other jobs in museums. Not to mention the many other jobs that are not related to museums in the art world, like art advising and working at auction houses. I had no idea about the wealth of opportunities that were available to me. And all of them require very different lines of experience and education.

My aim for the certificate is to provide students with more vocational guidance. I take them on field trips so they can see what's happening behind the scenes in museums and other community-serving institutions, and hopefully that helps them determine what internship might be best for them. We have class discussions about specific jobs and guest speakers visit. We also do projects in the museum and gallery practices class where each one is geared toward a different aspect of museum work. Students have the opportunity to think about, just at a very basic level: *Is this fun for me? Does filling out a condition report—that sort of detail-oriented work—look exciting and is it something that I can really get into?* Or am I more interested in talking to people about the artwork? In that case, maybe they're more interested in education than registration. So, that's my path and what's guiding what I'm doing right now.

I think it's helpful to be transparent about what you've gone through to get where you are and what might be necessary for them to get to where they want to go. As one of the exercises for the internship classes, I have students look for a job that they may want five years from now. I have them look really closely at the job description and everything else they would need to do and learn to actually get that job. It helps them think about what the job requires so that they are able to set themselves up. I also make sure to talk about the fact that it might be something that they want now, but as they go down the path, that may change. And that's okay, right? But it's important to be informed about where they might want to go.

Guerra: I think I would be classified primarily as a commercial gallery artist. I make art objects, which are paintings and drawings, and I exhibit them and sell them. Which is something that I know some of my students are interested in doing, too. I can provide them that hands-on guidance, if that's something that they choose to pursue. I'm an object maker. I use objects, drawings, and paintings to tell stories and communicate ideas. I've done illustration work professionally over the past decade. Though, I've kind of migrated more into gallery work because it's just easier to fit around my teaching responsibilities.

We talked a little bit about this, but the ideas and innovations that are coming out of other departments–because it goes back to this collaborative model–can be communicated very effectively and democratically through the stuff that we produce in the arts. As somebody who has an illustration background, I think being able to communicate ideas very clearly to a target audience that might not otherwise have access to information is a noble thing.

I'm super into comics, personally. Graphic novels. And I think there's a lot of opportunity there for sharing information with people about difficult subjects because comic books can be kind of disarming. I think the arts can be, as well. I mean, we can see how propaganda functions in our culture right now—and always—and you can see how effective images are in manipulating people's thoughts. I think it's really important that we have people who are very well-educated in how images work, so that they can hopefully be used for good. You know, whatever that means.

I think it's also hugely important for people in the sciences and people training in the sciences to have a strong humanities background. People who are going to become nurses or doctors or engineers or biologists or whatever; they need to be able to consider the human impact of their work. It's important that they're taught the skills necessary to frame their work within a wider context and understanding of power structures and their own humanity. I think there's a lot of things that are produced in the sciences that are harmful and deadly. And maybe an appreciation for the humanities and the kind of beauty of human beings might alter some of that decision making. Who knows? Maybe not. But that's what I would hope I can help do.

How can we make drawing, painting, photography and the aesthetic appreciation of the world that scientists are engaging with a part of their understanding of those spaces?

Woglom: Arts education has a kind of an interesting culture in that art educators often don't make their making practice central to their teaching practice. I have tried to kind of blend the two in a number of different ways. So, we do things like go to the art quad and build cardboard forts. But the idea is to make a weird thing and then see what happens because of that. Or we'll set up a bunch of tables during the *Campus Dialogue on Race* to have a space for folks to make art in relationship to what they're learning. I guess, to answer the question, making opportunities for folks to make art pieces ended up being my art practice, in a way. Making those moments where it feels like you can literally *make something* or that you have the agency to change things, I think, is a cool learning objective when going into classrooms and working with kids and making art. It's that, you know, we as a group can walk into this room at the beginning of the day. We don't have paper mâché puppets to do a puppet show with. But, at the end of a number of classes, we will. And you know that kind of affirms that we can make and change aspects of our lives. And then I also do a bunch of watercolor paintings and play music on the side.

Guevara: *Thinking about this upcoming semester, what kind of student work is planned that you hope will highlight the Polytechnic art experience?*

Potter: The students will get to collaborate on putting together an exhibition, which will be overseen by our gallery director Brittany Britton. Putting the work up on the walls is a really exciting process. Many students learn to hold a drill for the first time in that class. Putting on gloves and handling someone else's artwork, which is much different than how you would handle your own artwork. And then, of course, comes the excitement of opening the show they installed. No one else who's at the show realizes all the work that went into putting it on the walls. But the students know. When we walk into nice, clean gallery and museum space, we all think: *oh, it looks pretty easy*, right? But we don't think of the hours and hours of work that goes in, even just in the install part. Students always talk about that: "I had no idea that all of this was involved!" It's so exciting to see how proud they are by the end.

I think some of the students walk away thinking: *oh my gosh, there's so much more math involved than I ever imagined would come with hanging all these works straight*. And some of them get really excited by it. And even for the ones who realize: *wow, this work isn't for me*, or *this aspect of the work isn't for me*, that in and of itself is a really good learning experience, right?

Woglom: There's our service-learning course—which is the one that I'll be teaching in the spring—we've planned it to be a two-part course. In the fall, students wrote a piece of curriculum that they will then enact at a space. So, this semester at Pacific Union Elementary, students will be going to create paper mâché masks and then do a theatrical piece with another artist who's going to come and teach them a traditional dance structure. That's going to be super fun.

We also have a group that's working with a local high school to do an enormous mural on the exterior of their building. That'll be going on for the semester. We have a group that's teaching ceramics over at Arcata Arts Institute, which will be exciting. Then, there's a group that is working with our Prison Arts Collective, which is a partnership we have with San Diego State University, where we go into Pelican Bay State Prison and do arts programming there. That group has been working on curriculum for them. We're setting up a show that will be going up in the second week of February.[d] We're going to try to Zoom in some of the dudes from the prison, and someone who is recently released is going to come and talk about his work.

***Guevara:** Can you reflect on elements of art education and art worlds that stand apart from the polytechnic model. What do you hope the campus and broader community understand about these parts of the art worlds and their significance for students and the community?*

Potter: I don't see a stand apart–so much as an alignment. Like with everything we've already said about hands-on work. The hands-on work that the students are getting is really valuable, not only in terms of them learning what types of paths they want to pursue, but also to make them more desirable in the job market. I was just talking to Nicole Jean Hill, our department chair, who was at a photography conference. There was a panel at the photography conference specifically about jobs in photography in museums, such as photographing collections and exhibitions. And some of those museum representatives were there presenting specifically because they were having a hard time finding new employees who have a specific vocational background related to museums and photography. That's something really special that our program can offer that isn't being offered at as many institutions.

Also, museum studies is a growing field. It's exciting to see that more museum studies programs and certificates are being offered. When I was applying for a master's in museum studies, there were very few programs. But now the field is becoming so much larger because these graduates of the programs go on to serve so many more institutions than just museums. Think about all the historic houses. The national park system in the U.S. is one of the biggest museum systems in the world. So, there's a lot of work to be done and hopefully students are excited to pursue museum work.

Guerra: My experience so far has, I think, been that the definition of "polytechnic model" shifts depending on who you're talking to. And I think maybe there's a concern that there is an extreme focus on career education. Maybe a focus on prepping people for corporate jobs. Of course, people have to eat. I understand that. We are a part of this world, for better or worse, and people have to survive. But I think that maybe things that don't immediately generate profit can still be valuable. Without getting too deep into it, not everything has to immediately be a profit generating endeavor. Human beings are about more than just laboring and producing money. That's something that we have to kind of keep in mind for the reasons we discussed about needing people to have a humanities and art background. It's an important thing.

Cal Poly Humboldt Art Education students, along with students from Redwood Coast Montessori High School and local community members, paint a 20'x250' mural on the northern exterior wall of the high school's building at the corner of 8th and K St. in Arcata. Photo credit: Sonia Sanchez

Woglom: To be honest, my brain kind of goes towards connections. And I think I would actually have a hard time thinking about what couldn't be a possibility. The more I hear about things that are happening, my brain immediately goes to like: *How could we do something aesthetic with that space? How could we engage with artistic understanding through that space?*

And, I don't know, I honestly think we can do all of it. That also may be, you know, potentially overzealous. But, I mean, why not? I'm so proud to be involved with these folks in the Department. They're just so committed and engaged and constant in their working to make cool things happen that it's just like, I don't know, it blows my mind every day.

Putting Words Into Action

HJSR is no stranger to the promotion and inclusion of art, both in our pages and in multimedia formats hosted on our website. To quote *HJSR*'s editorial board president, Mary Virnoche, "We feel strongly about the imperative to integrate the arts and humanities across conversations, texts, and images that engage the polytechnic institutional identity." In the early stages of bringing this issue together, my team of editors and I contemplated our options for the inclusion of artwork that remained relevant to our theme. Alderson, a faculty member in the Art + Film Department, was immediately enthusiastic about collaborating with me to create a unique space for Art + Film faculty voices in this issue.

As mentioned previously, the interviews with Guerra, Potter, and Woglom were held in early 2023. At that point, none of the student photography displayed here had yet been produced. As discussions between Alderson and myself progressed, we concluded that to best represent the hands-on, experiential-learning occurring in the Art + Film Department, we wanted work from students that was both timely and multidisciplinary. At the time of publication, the art photos are a scant few weeks old, produced by students still basking in the glow of another finished semester.

Alderson then reached out to Art + Film Department Chair Nicole Jean Hill, the faculty member behind the production of student photography featured in this issue. The following excerpts of the conversation between Alderson and Hill further detail the Department's continued commitment to the polytechnic model of education expressed by Guerra, Potter, and Woglom.

Alderson: *Could you tell me a little bit about yourself and how you found your way into the art world?*

Hill: I'm the chair of the Department of Art + Film at Cal Poly Humboldt, and my main discipline is photography. I found my way into the art world through being interested in journalism. I wanted to be a photojournalist. But then I discovered the complexity and excitement that was the fine art world, and specifically, photography within the fine art world. So, I ended up pursuing both a bachelor's degree and a master's degree in studio art. I've been a practicing artist and educator ever since.

Alderson: *Can you describe your work as a member of the Art + Film faculty at Cal Poly Humboldt?*

Hill: In my teaching capacity, I have been leading the photography area for many years. I was instrumental in expanding the photography curriculum to having digital and color photography instead of only analog black and white. I've spent a lot of time refining that curriculum, working with a lot of amazing lecturers to implement it, and trying to keep photography relevant to both the fine art world and to the world of commercial photography that a lot of our students find themselves in after graduation.

Alderson: *As an art faculty member in the Department, what's the collective vision for growth and development that have roots in the polytechnic conversations?*

Hill: We really look at the polytech expansion as a way to continually improve and refine our existing majors in art education, in art history and museum studies, and in studio art. I think that one of the key parts to our art program—and one of the strengths that I feel like the emergence of the polytech concept fits together well with—is the idea that we are constantly thinking about our program as: *Where do we want the students to be at when they graduate and what avenues will they take beyond graduation?* So, we've back-engineered our programs to get students thinking about careers and to get them involved in experiences on campus that help to build up their resumes. And, basically, I want them to have a much different experience than I had in art school. I was never asked to think about those things until I was already out of there. It's really important to me, then, to build all of this directly into the curriculum. So, the polytech seems like an opportunity to continue to refine those things that we are doing really well, and to have the resources and fa-

We've back-engineered our programs to get students thinking about careers and to get them involved in experiences on campus that help to build up their resumes.

Art+Film student Cita Hunter sketches in the Cal Poly Humboldt Greenhouse. Photo credit: Alisa Ibyadullin

cilities to communicate that to students as they're working their way through the program.

In terms of expanding things in new areas, I think that the timing of us merging with the film program has been an amazing opportunity for us to think about the ways that we intersect with this other discipline. And from there we can build a new major that solidifies the intersection between art and filmmaking in the realm of media art. We're accomplishing this by creating a new BFA in media arts—although that's a working title—that merges the strengths of our art program with the strengths of our film program. Ultimately, it's a BFA in media arts with emphasis areas in film production, interactive media, and digital art. Becoming a polytechnic allows us to both expand our existing degree programs and to offer additional professionalized degree programs for our students.

Alderson: *When we were in conversation about this project for the journal, we thought that in terms of highlighting what our department does, the idea of having students provide the visuals for the essay was an exciting idea. How do the images that are accompanying this article relate to the kinds of activities that students already do in our photo classes on campus?*

Hill: What we're trying to do in the photo curriculum is give the students the technical skills to use camera gear from cameras to lighting to printers, etcetera. And to be able to creatively problem solve within certain parameters, but still have the freedom to express themselves, learn to deal with issues of composition and design, and to be able to effectively communicate a story within a single frame. So, we built the curriculum starting with fundamentals—like learning the tools—first. Then, in the intermediate and upper division courses, we get them to build a portfolio that has their own unique style of photography where they're using the tools in a very professional way. The idea is that we're giving them assignments that replicate the work they might actually have in the world once they're working as freelance photographers or as photo editors or what have you.

This particular article and the accompanying photographs are a really good example of a kind of editorial assignment that you might get for a major publication where they are looking for an environmental portrait of an individual that communicates a certain time, place, and activity that doesn't feel overly staged. It feels, you know, kind of natural, and has a certain level of technical quality that would match what the publication is presenting in their other work. So, we gave these students multiple assignments that deal with environmental portraiture.

I feel that this was a real opportunity for the students. I mean, normally they get to pick their own subjects. Usually, they'll photograph their coworkers, their friends, or their family. And when they do those kinds of projects, the real challenge—what I always tell my students—is that you don't take a photograph, you solve the photograph. So, if they're going to photograph their friend in their apartment because they got an assignment from me, then it might seem really comfortable. But it's up to them to figure it out in that space. Does the student have a big window where a bunch of natural light is coming in? Do they have a visually distracting poster on the wall? Do they have worn shag carpeting? What are the people wearing? These are all things that they have, potentially, no control over but still have to solve into a photograph. This opportunity to photograph for the journal is another extension of that.

So, we've given the students this assignment that's similar to what they might get from a magazine. Something like: *We need you to go and photograph this artist. We want a picture of them in this type of location. Go and figure it out.* They're faced with solving that photographic problem, while also being able to be responsive to the needs of the publication. When photographers work for publications, they are often given certain parameters, like: *We need X number of photographs. We need a combination of close-up shots and distance shots. We need verticals. We need horizontals.* And so, the students were asked to replicate that previous experience they had with the assignment that was in their comfort zone—you know, with their friends and family—but now it's an assignment where they're in control of even fewer variables. And then, hopefully, they've been prepared to solve that visual problem. In this particular situation, we asked the students to work with lights, and to work with their digital cameras in a way that is similar to their assignments, but again, fits the specific needs of *HJSR*.

Alderson: *You gave them some specific targets, in terms of the number of photographs or the kinds of photographs, that are similar to what would happen if they were being hired by a publication. I know there are lots of photo faculty that also do work like that for major publications. Was the goal to make it replicate that kind of an assignment?*

Hill: Yeah, you would normally get a shot list from a client, whether that be a publication or something else. And they usually don't know exactly how things are going to fit in. They don't yet understand if they're going to need verticals or horizontals, or if they're going to need something that can really read at a small scale, or if it's going to be a cover photograph. So, you're usually given these parameters to allow the publication a lot of options to choose from. They need a lot of options so that the photo editors and designers can decide what to use. It was kind of the same thing with the students, you know. We'd say, "Don't, just give us one." And that's the difference, right? In an art class, it might be: Go and take a portrait of someone, and then hang on the wall the one that you think is best. But out in the world, you're oftentimes asked to give a buffet of options. You have no control over the final one that's picked. So, you want to try to do your best work in the selection that you're giving them.

Alderson: *One of the things that I was impressed about, in terms of the conversations that we had with the students, was that one of the first questions they asked was, "Is it going to be color or black and white?" That's a consideration that a lot of people don't think about. It's not quite as simple as either in color or in black and white. It changes everything about the visual of it.*

Hill: Yeah, totally. And afterwards, when we were looking at the photographs, the students would realize that maybe one photo would read a lot better in black and white, while another would read better in color. And then, for some publications the cover might be in color, but the inside photographs are black and white. You have to attempt to balance those things out and try to pre-visualize as you go. Again, it's good for the photographers to provide a lot of different options because what might actually work well isn't always super obvious as you're shooting. We try to think about what will work in black and white or what color might be too distracting. As image makers, we must train ourselves to have a little sense of that. But you really have to just pre-visualize as best you can, and then be prepared with a lot of potential options to hand over to the client.

Alderson: *Can you share any thoughts about the images that the students produced?*

Hill: Well, one of the amazing things about the magic of photography is that it's form and content at the same time. I was just really excited to see the way that they problem solved in, you know, really overwhelming environments like the greenhouse, the gallery, and the wildlife building. From a formal side of things, I think they did a really great job in the selections that they chose and in understanding what makes a clear image without a lot of distraction, and that can work on different scales of print size, whether it be a small image or a cover image. I think they did a really good job of that. But it's hard for me to separate my excitement, just in terms of the content. To see all these awesome students doing all this cool stuff and all this hands-on learning, it's just exciting! I would have been excited about the photographs, even if they hadn't visually problem-solved the shoots as well as they did. But I think that they did a really nice job of merging the form and content together with the photos they submitted to us.

Alderson: *When we were talking through this idea of the Art + Film Department being involved in this issue of the journal, there were a lot of different conversations*

Art 356: Museum and Gallery Practices students install the 2023 Art Graduates Exhibition at the Reese Bullen Gallery. Photo credit: Sonia Sanchez

about what it might look like. And what we have—these conversations with our faculty related to these particular areas—that actually comes out of a conversation that I had with you. We were talking about what we already do that's related to hands-on learning. And you brought up these three areas, right? Scientific illustration, museum and gallery practices, and art education. We then came up with the idea of the photo piece as replicating real world job experience. But the whole idea of the faculty being interviewed, along with this whole structure comes from a point that you made about the fact that our department already does all this hands-on learning, and these are the particular areas where that's most visible and obvious. Is there anything else about that that you'd like to say?

Hill: I guess it just kind of goes back to all the folks who have been asking us, the art faculty, about how we feel regarding the polytech. Asking, "Aren't you nervous? Aren't you scared?" But to me, I mean, I've already been familiar with some other polytech institutions. And I know that polytech is about hands-on learning. When that announcement came out, I was like, "Yeah, no, of course, that's what we are." Becoming a polytechnic is just recognizing what we were already doing because we're an Art + Film Department. It's not a stretch to think about how we fit within the polytech at all. I do understand the concerns that other disciplines might have in the humanities, even though I know there's so many different ways that they are also hands-on. But for us, in all of our majors—even in art history, which seems like it would be the least polytech of all of our disciplines—it's so incredibly hands-on. The students are working in their classes in the art history museum studies program like how the students in the studio classes are working. They're engaged in exhibitions. They're engaged in handling objects.

I just think that it's a really good fit for us, and I think it's really exciting. It's aligning with our strengths, and that allows us to expand those strengths even more.

Becoming a polytechnic is just recognizing what we were already doing because we're an Art + Film Department. It's not a stretch to think about how we fit within the polytech at all.

References

California State University. 2022. "What is a Polytechnic University?" The California State University. Retrieved March 22, 2023. https://www.calstate.edu/csu-system/news/Pages/a-polytechnic-transformation.aspx#:~:text=Polytechnic%20Defined,%2C%20technology%2Dbased%20skills%20training.

Creative Coast. 2022. Creative Coast Landing Page. Cal Poly Humboldt. Retrieved April 16, 2023. https://creativecoast.humboldt.edu/.

Deleuze and Guattari. 1987. *A Thousand Plateaus: Capitalism and Schizophrenia*. Translated by B. Massumi. Minneapolis: University of Minnesota Press.

Jones, Carolyn. 2022. "What Prop. 28 Ffunding will Mmean for Aarts Eeducation in California." EdSource. Retrieved March 28, 2023. https://edsource.org/2022/what-prop-28-funding-will-mean-for-arts-education-in-california/682054.

Tanner, William R. 1993. *A View from the Hill: A History of Humboldt State University*. Arcata, CA: University Graphic Services.

a. As part of Cal Poly Humboldt's PBLC programs, which are offered to first-year students in a variety of majors, students begin their college journey with Summer Immersion. Creative Coast's Immersion "explores connections to the local region, as well as the local art community" (Creative Coast). Learn more about the Cal Poly Humboldt's PBLC programs here: https://pblc.humboldt.edu/learning-communities-list

b. The mural, entitled "We Been Healing That Way" can be viewed in person on the westward facing wall of the Pacific Outfitters/Hatchet House building in downtown Arcata, or online at the REBOUND website: https://www.r-e-b-o-u-n-d.net/tripp01

c. Find out more about ShakeAlert and experience their visual style here: https://www.shakealert.org/

d. Select photography and more information on the exhibition, More Than A Number, and its forthcoming virtual tour can be found online at the El Leñador website: https://www.el-lenadornews.com/2023/02/15/artists-from-pelican-bay-state-prison-open-exhibit-at-cal-poly-humboldt-reese-bullen-gallery-on-feb-16/

HJSR
50TH ANNIVERSARY EDITION

Polytech to PolyTEK: Traditional Ecological Knowledge, Indigenous Science, and the Future Forward Polytechnic University

Cutcha Risling Baldy *<csb357@humboldt.edu> (Cal Poly Humboldt),* ***Kaitlin P. Reed*** *(Cal Poly Humboldt),* ***Kayla Begay*** *(Cal Poly Humboldt)*

Abstract

It is clear from Cal Poly Humboldt's Polytechnic Prospectus that Traditional Ecological Knowledge (TEK) and Indigenous communities are key parts of what elevates Humboldt's development of a polytechnic university for the next century. The prospectus demonstrates Humboldt's proposed framework for a different comprehensive polytechnic "will also be informed by Indigenous communities and ways of knowing, as many Native peoples have lived sustainably in their places since time immemorial" (19). There are many considerations when engaging with TEK, especially around sustainable use. It is also important that engagement with TEK and Indigenous science not only center knowledge sharing, but also how departments, programs, and colleges are dedicated to upholding sovereignty and self-determination and working to empower Indigenous students, communities, and ongoing projects of land return, environmental justice, and education. This article will discuss the role of Native American Studies in building decolonial frameworks for a new polytechnic—polytech to PolyTEK. The article explores the history of cultural knowledge exploitation, Humboldt Native programs and initiatives; the resurgence of Indigenous science and knowledges, and new interdisciplinary initiatives at Humboldt that value NAS as a partner to building polytechnic programming.

Humboldt is positioned to offer a cutting edge and unrivaled polytechnic experience to current and future students. Indigenous knowledge systems are especially important and appropriate to consider in the development of a polytechnic institute because Indigenous knowledges are fundamentally interdisciplinary and applied. Indigenous knowledges are also at the forefront of cutting-edge research interventions in the sciences and western academic institutions. When we talk about or propose "decolonizing" curriculum or higher education we must build this from Indigenous frameworks with Indigenous Peoples at the center of our academic vision and planning.

Keywords: Traditional Ecological Knowledge, Native American Studies, indigenous science, sustainability, decolonization, STEM education

On December 8, 2021, esteemed scholar Robin Wall Kimmerer (Citizen of Potawatomi Nation) joined Cal Poly Humboldt students, faculty, and administrators for a virtual discussion titled "'Poly-TEK?' A Discussion with Distinguished Author Robin Wall Kimmerer," about how Humboldt can ethically and sustainably incorporate Indigenous sciences and Traditional Ecological Knowledge (TEK) into the development of a polytechnic institution. Cal Poly Humboldt alumni Marlene' Dusek (Payómkawichum, Cupa, and Kumeyaay) opened the panel discussion with a question: "Dr. Kimmerer, what would your recommendations and ideas be for the first steps that HSU[1] and any other settler-colonial university should be taking to center Indigenous knowledges and TEK, and, how do you think they genuinely begin to repair the relationship with the land and the peoples, whose stolen land they sit upon and occupy, without just re-duplicating another broken treaty?"

Dr. Kimmerer responded by referring to the ongoing work of the Native American Studies (NAS) Department at Cal Poly Humboldt to engage in faculty learning circles about TEK as "inspirational" adding:

> I think the first thing that we do is tell the truth, when we are dedicated, as we all are, as academics, as students, and teachers, and I hope, and I know we're all those things simultaneously, students and teachers, that we have to have a dedication to telling the truth, which really means telling the truths. But then we say, well, what do we do about that? How do we act upon our recognition of historical wounds? And, how do we enact our role as healers? Because that's what we all are, as teachers and students, we are healers of land, and healers of relationship. So, that telling of the truth needs to go into every corner of the curriculum… because, in my experience, oftentimes, these issues of cultural justice and land justice tend to be siloed. They tend to occur in the margins, right?

Dr. Kimmerer asks that we think critically about how we build cultural and land justice into what it means to be a "polytechnic" university. What does it mean to be a 21st Century polytechnic university that is occupying dispossessed Indigenous lands and continues to benefit from Indigenous lands and resources? How do we at Cal Poly Humboldt engage with and foreground an ethical praxis of being in place?

It is clear from Cal Poly Humboldt's Polytechnic Prospectus that TEK and Indigenous communities are key parts of what elevates Humboldt's development of a polytechnic university for the next century. The prospectus demonstrates Humboldt's proposed framework for a comprehensive polytechnic that "will also be informed by Indigenous communities and ways of knowing, as many Native peoples have lived sustainably in their places since time immemorial" (HSU 2021:19).

Cal Poly Humboldt is the northernmost California State University (CSU) campus. The Humboldt campus is located in the traditional homelands and unceded territory of the Wiyot people which includes the Blue Lake Rancheria, Bear River Band of Rohnerville Rancheria and Wiyot Tribe. Humboldt County is home to the three largest Tribal Nations in the state of California—the Yurok, Karuk and Hoopa Valley Tribes. Cal Poly Humboldt also enrolls one of the largest percentages of Native American students in the California State University system. We are the only four-year university in our rural region and though we are among several thriving Tribal Nations, there are no Tribal colleges within our service area. In addition to Humboldt having a well-earned reputation in the fields of sciences and natural resources, it is also home to some of the longest standing Native and Tribal Programs within the CSU and University of California (UC) systems, including the NAS Department, the ITEPP (Native American Center for Academic Excellence) and INRSEP (Indian Natural Resources, Science and Engineering Program + Diversity in STEM) programs. Humboldt's local admission and service area is also home to many of the Tribal Nations who are leading the way in managing and co-managing their

1. This panel was held before HSU officially became Cal Poly Humboldt. References throughout the article in quotes or previously published articles or documents may refer to now Cal Poly Humboldt as "Humboldt State University" or "HSU."

traditional lands, waterways, airways and natural resources with methods that include traditional values and traditional science coupled with western science practices. Ongoing research projects and educational programs with the Tribal Nations of our service area have already made Cal Poly Humboldt an important collaborating institution with Tribal Nations in this region.

NAS is a discipline that has existed for Native peoples for time immemorial. For some, NAS emerges during the "intellectual and political ferment of the late 1960s and early 1970s" enveloped within the Red Power Movement, the development and activism surrounding Ethnic Studies in the academy, and the continued struggle for Indigenous self-determination (Kidwell and Velie 2005:3). However, the development of NAS has roots and connections to a long history of education and intellectual disciplines of Native peoples that stretch back for time immemorial. NAS in the academy, in part, developed out of an interest and need to shift academic research and disciplines away from the study "of" Native Americans to the study "with" and "for" Native peoples and communities. NAS was not necessarily welcomed into the academy, but instead was the result of numerous protests and activism by students, community leaders, faculty and staff at various universities across the United States (Cook-Lynn 1997). The development of programs, projects, majors and minors, certificates and other areas of academic development and community-facing projects should invite and include NAS as partners and encourage the support and involvement of a discipline that represents the vision of Tribal Nations and peoples. It is clear that "place-based learning" cannot occur without NAS curriculum, as the lands the university occupies are Indigenous lands with ongoing critical Indigenous connections across organizations and Tribal Nations. It is also important that engagement with TEK and Indigenous science not only center knowledge sharing, but also how departments, programs, colleges and the university are dedicated to upholding sovereignty and self-determination and working to empower Indigenous students, communities, and ongoing projects of land return, environmental justice, and education.

This article will discuss the role of Native American Studies in building decolonial frameworks for a new polytechnic—**polytech to PolyTEK**. The article explores the history of cultural knowledge exploitation, Humboldt Native programs and initiatives; the resurgence of Indigenous science and knowledges, and new interdisciplinary initiatives at Humboldt that value NAS as a partner to building polytechnic programming.

Traditional Ecological Knowledge & Indigenous Science

On November 15, 2021 the White House Office of Science and Technology Policy released a national-level memorandum regarding the importance of TEK and collaboration with Tribal Nations. In this memo they recognize TEK as a contributor to the "scientific, technical, social, and economic advancements of the United States and to our collective understanding of the natural world" (White House 2021). It also stated that TEK is essential, "for developing comprehensive climate adaptation and natural resource management strategies" (White House 2021). Indigenous science is at the forefront of how climate scientists are thinking about our shared futures and it is clear that students who have a foundation in both Indigenous science and western science perspectives are more prepared for their careers.

Traditional Ecological Knowledge (TEK) or Indigenous Environmental Science (IES) is broadly defined as encompassing millennia of environmental knowledge passed down through generations through observation and practice of spiritual understandings and interactions with the environment (Anderson 2005). TEK is modern and adaptive and may have different meanings to different Indigenous groups. TEK is foundationally centered on interdisciplinary, interconnected approaches to scientific interventions. TEK is also necessarily regionally and Tribal Nation specific and cannot be singularly defined or applied but instead requires thoughtful engagement with place. Unlike western conceptions of conservation, preservation, and natural resource management where humans are considered outside of nature, TEK considers that humans are integral to functioning ecosystems and are a part of nature (Pierotti and Wildcat 2000). According to Dr. Jessica Hernandez (Zapotec/Maya Ch'orti') the foregrounding of western conservation and environmentalism in western science fundamentally ignores how "conservation is a Western construct that was created as a result of settlers overexploiting Indigenous lands, natural resources, and depleting entire ecosystems" (Hernandez 2022:72). Hernandez's book *Fresh Banana Leaves: Healing Indigenous Landscapes through Indigenous Science* offers a powerful critique of the ways in which settler colonial ideology has always, and continues to, inform environmental knowledge and practice within the United States and Latin America. Within her text, Hernandez employs the concept of ecocolonialism, which she defines as:

> ...the altering of our environments and landscapes due to colonization of Indigenous lands and the paradigms that are upheld to grant settlers (white people) the power to continue managing our environments and landscapes. *Every environmental scientist, researcher, policy maker, and anyone else who ever took a decision over our environments without consulting the Indigenous peoples of those environments has practiced some form of ecocolonization* (Hernandez 2022:42 emphasis added).

Western ideological conceptions of terra *nullius*, or nobody's land, marked the landscape of North America as open for removal and dispossession of Indigenous peoples. Notions of manifest destiny engendered a moral justification to colonize and develop territories. The western concept of wilderness aligns with *terra nullius* and also erases Native presence and ecological management practices. Wilderness, however, had to be *created* (Blackburn and Anderson 1993; Cronon 1996; Spence 1999). As Anderson (2005) argues, California was far from an idealized wilderness when invaders first arrived—what the Spanish and the miners stumbled upon was actually a well-tended landscape.

TEK has been described as a "third alternative" for science, conservation, and sustainability work, where Pierotti and Wildcat (2000) argue that "TEK is inherently multidisciplinary in that it links the human and the nonhuman, and is the basis not only for Indigenous concepts of nature, but also for concepts of Indigenous politics and ethics. This multidisciplinary aspect suggests that TEK may be useful in resolving conflicts involving a variety of stakeholders and interest groups in controversies over natural resource use, animal rights, and conservation" (1333). TEK is also an imperfect term applied to a complex system of knowledge that engages science, culture, spirituality, political, and social praxis. Reframing Indigenous science and knowledges as TEK may separate the methodological interventions of Indigenous science from western science or more widely accepted disciplinary labels like "Ecology" or "Biology" or "Chemistry" or "Zoology." There is no doubt that Indigenous TEK includes interventions and methodologies of ecology, biology, chemistry, astronomy, zoology, and other fields, yet it is rare to see Indigenous TEK valued or centered as part of these western scientific disciplines. Hernandez discusses how the terminology of TEK can "historicize" Native peoples and their knowledges. Hernandez advocates for the use of "Indigenous Science" defined as "Indigenous voices, perspectives and lived experiences... [which] embody our ways of knowing that are rooted from ancestral knowledge and valid sciences" (Hernandez 2022:13). Dr. Kimmerer expands on a critique of western terminologies in her December 2021 discussion at Cal Poly Humboldt:

> When we start talking about rivers and berries and trees and mountains, as natural resources, well, we know that's violent language. That's damaging language to say that these lands are not our relatives. These lands are not our pharmacy, and our library, and our elder. They're natural resources, they're commodities, which are only valuable when they are extracted. That's really damaging kind of language. ..Well what's the alternative? You can have a department of natural resources, or you could have a department of earthly gifts. Just that language changes everything, doesn't it? It changes everything, from thinking about extraction and rights, to thinking about responsibility. Language has tremendous power.

Intervening in the discourse surrounding TEK to center Indigenous language of place has been an important way to demonstrate the diversity of Indigenous Science and TEK. Wildlife biologist Dr. Seafha Ramos describes "Yurok TEK" through Yurok language as hlkelonah ue meygeytohl ('to take care of the Earth'). Yurok TEK is expressed as striving to create and maintain balance between people, wildlife, and the environment via "physical and spiritual management in tandem" (Ramos 2019:86).

The development of a "polytechnic" university at Cal Poly Humboldt can be a cutting-edge and relevant pursuit needed to address some of the most pressing environmental issues of our time. Daniel Wildcat (Yuchi/Muscogee) (2010) argues that Indigenous Sciences provide key foundational concepts and approaches that demonstrate their ongoing contributions and relevance to science and technology. First, Indigenous knowledge production is thousands upon thousands of years old, much longer than the creation of and knowledge gathering of most western academic disciplines. As Dr. Kaitlin Reed (Yurok) argues "California Indians have been here since time immemorial and we have had a lot of time to experiment, to figure out what works and what doesn't. Yet despite the literally tens of thousands of years California Indians have been studying their environments, Western scientists—who have been in California for 170 years maximum—assume they hold intellectual superi

Wildcat (2010) also argues that Indigenous knowledges ask students, scholars, and practitioners to connect through a spiritual and cultural interrelationship with land. For example, climate change is a cultural issue and will necessitate a change in worldview. Dr. Kimmerer adds to this in the context of understanding how a change in culture will necessitate changes in how we approach science and technology: ority on the functions of our ecosystems" (Reed 2023:6).

> Hypothesis testing, and the criteria that we have for what constitutes valid knowledge, are all really important when the question is true/false, and that's essentially what hypothesis testing is right? And we very carefully separate the observer and the observed, so that these questions of faith and subjectivity don't influence our answers to a true/false question. I honor that. That's kind of the canon of western science. ...What if the question isn't a true/false question? What if it's a right/wrong question? In that case, our objective, materialist true/false hypothesis testing needs something else. It needs values inserted into it. It needs subjectivity. ...And, those questions that require an examination of our values, and our assumptions of what are our responsibilities as human people. Science, western science, can't answer that. We've explicitly cut ourselves off from incorporation of values, for good reason, because we want to answer true/false questions. But I think we have to honor the incorporation of subject knowledge, subjective knowledge, and values and belief systems, and spiritual ways of understanding spiritual responsibilities, in the realm of more complex questions about, what should we do? How is it that we should live?

Cal Poly Humboldt, Traditional Ecological Knowledge, and Indigenous Science

Cal Poly Humboldt has a unique opportunity to be a leader both statewide and nationally in implementing Indigenous perspectives across university curricula and programming. The region has a significant Native American population, including 12 Tribal Nations within Humboldt County alone. In California, the percentage of Native Americans or Alaskan Natives is reported at 3.6%, one of the highest percentages in the nation, whereas in Humboldt County that number is 11.3%. Del Norte County, located just 40 miles north of Cal Poly Humboldt, is approximately 14.4%. NAS has expanded their faculty over the past five years and has been integral to ongoing partnerships with the College of Natural Resources and Sciences to integrate and engage with TEK, Indigenous Science, and decolonial frameworks for educational curriculum that addresses the most pressing needs of the future workforce in STEM. NAS has partnered with the Forestry Department to develop a Tribal Forestry Concentration and with the School of Engineering to create a Masters of Engineering and Community Practice which includes a rigorous curriculum in NAS and Indigenous Science/Traditional Ecological Knowledge.

In the Cal Poly Polytechnic Prospectus, Tribes are mentioned over 75 times (HSU 2021). In total, the prospectus engages with terms like "Tribe," "Indigenous," "Native American," "Indian," and "Traditional Ecological Knowledge" 198 times. The Polytechnic Self Study quotes a 1989 consultant report titled "American Indians and Humboldt State" written by leading Indigenous scholar Vine Deloria, Jr. (Standing Rock Sioux). Deloria's report is an essential document highlighting the role of NAS on our campus. But he was clear in his recommendations at the time that there needed to be much more support of the work that Native programs were doing to elevate the University. Deloria, Jr. recognized this in his report when he wrote: "The most impressive part of my visit to Humboldt State was to see how much the existing staff has accomplished with minimal to barely existing resources" (Deloria 1989:9). The accomplishments that impressed Deloria did not happen due to the goodwill of the university, but were hard won victories fought for by students, faculty, and staff that continued well after Deloria's visit in the late 1980s. For example, the Behavioral & Social Sciences (BSS) Building on campus was constructed in 2007 and included the Native Forum as a space designated for Native American programming on campus. The design of the Native Forum on the first floor of the BSS is modeled after a traditional house for Northwest California Tribes. The official name of the space is the "Native American Forum" and has been since the opening of the BSS Building in 2007. In January of 2015 Humboldt State students occupied the Native Forum with specific concerns about many issues, but one of them was the designation, signage and support for the Native Forum and Native programs on campus. Students expressed frustration that while the BSS building had originally included an entire first floor that was dedicated to Native programs, over the years these spaces had lost any connection to Native peoples. Because of this student activism the Native Forum was re-dedicated and officially named as the "Native Forum." In October 2019, NAS was able to secure funding and remodel the lobby outside of the Native Forum, something that was appreciated by Tribal peoples and students because it brought in a clear Indigenous presence to the space. For Cal Poly Humboldt to prioritize Tribal knowledge systems and community partnerships in the development of a polytechnic institute, Humboldt needs to put those commitments to action.

On April 29, 2021, the Council of American Indian Faculty and Staff (CAIFS) released the "Statement on PolyTech University Development and Indigenous Knowledge/Indigenous Science/ Traditional Ecological Knowledge" (included in full as an appendix), written as a collaborative statement from the Native faculty and staff at Cal Poly Humboldt. CAIFS is

> an advisory council that includes members from many of our Humboldt Departments, Programs and Colleges across campus. This multi-tribal , interdisciplinary group meets regularly to discuss complex issues facing our American Indian communities with a focus on campus and regional matters (CAIFS 2021:1).

The document was submitted to the Polytechnic Prospectus working groups in order to provide a guiding document to help ensure that tribal voices are not only included but become a part of the foundational fabric and ongoing leadership of this opportunity in front of us" (1-2). The document includes "Guidelines for TEK engagement and partnerships with Indigenous communities" including:

> To be a leading polytechnic means naming and engaging with TEK across this campus and it is important that courses that center Indigenous perspectives, (such as Native American Studies, American Indian Education, Decolonizing Social Work, etc.) not be treated as an "elective" to majors but instead as essential courses to our campus curriculum. Indigenous knowledge systems are relevant to every field on campus, ranging from health, art, business, law, governance, sciences, etc. (7).

There are many examples of successful integration of Indigenous perspectives on the Humboldt campus to look to as models for building a PolyTEK university. Rroulou'sik ("rising up" in Wiyot) was a 3-year funded National Science Foundation Research Experience for Undergraduates Project that encouraged Native students to work on "natural resource management research topics relevant to Indigenous peoples." The Place-Based Learning Communities (PBLCs) in the College of Natural Resources and Sciences include NAS curriculum as an integral part of the curriculum. Since fall 2020, Cal Poly Humboldt has been the only CSU where over 80% of the incoming first-year students who are majoring in the sciences also take a NAS course. Johnson et al. (2020) found that students who participated in PBLCs "had stronger sense of belonging, improved academic performance, and increased first-year persistence... Equity gaps were narrowed in first year GPA, pass rates in several gateway STEM courses, and in first year persistence at the institution and in a STEM major specifically" (522). This demonstrates both the significance and necessity of ethically integrating Indigenous perspectives and knowledge systems in STEM education.

As part of the PBLCs, incoming first year students experience a week of summer immersion prior to the start of classes. The NAS Department has facilitated interactions between students and Tribal leaders in the local community. For example, the Among Giants PBLC, which focuses on mammal and plant biodiversity within the coastal redwood ecosystem, summer immersion experience has included a welcome to Wiyot territory at Cal Poly Humboldt by Wiyot Tribal Chair Ted Hernandez. Mr. Hernandez offered the students a prayer, taught students about Wiyot history and worldviews, and with the help of Rachel Sundberg (Ner-er-ner, or 'Coastal Yurok,' enrolled at Trinidad Rancheria), showed students regalia, songs, and dances. The following day, Among Giants students traveled to Prairie Creek where they were welcomed by Yurok Tribal leaders Susan Masten and Linda Cooley. Susan Masten is a former Yurok Tribal Chairperson and former President of the National Council of American Indians. Linda Cooley is the Deputy Director of the Yurok Tribal Economic Development Corporation. Both leaders shared the history of the Yurok Tribe, Yurok Tribal worldviews and natural resource management practices.

As part of the PBLC experience, the NAS Department—in partnership with the PBLCs and ITEPP—created two educational videos to aid in student learning that features NAS faculty, local Tribal leaders, and several Indigenous alumni. One of the videos, entitled *History of Native California*,[2] is the most viewed video on the Humboldt PBLC YouTube channel by a significant margin. As of April 2023, this video had over 135,000 views. This video has also been shared in a variety of classes, academic conferences, featured on local news stations, and even screened for the California State Legislature. The other video, entitled *TEK & PBLCs*,[3] demonstrates the significance and necessity of TEK within the natural resources sciences. The partnerships created between NAS and science departments in the College of Natural Resources and Sciences while developing and implementing the PBLCs is unprecedented and has created public discussion and outreach to NAS from universities throughout the nation and globe interested in learning more about this curricular milestone. The NAS Department welcomes these conversations, and is proud to represent Cal Poly Humboldt on the international stage as a leader in the field. Additionally, the NAS Department is currently developing a third video focused on TEK in the humanities and social sciences to demonstrate that the applicability of TEK is not limited to natural resource sciences, but every field of study at Cal Poly Humboldt.

NAS has also developed and implemented community-informed department-led initiatives that contribute to the integration of TEK in higher education. This includes the establishment of the Rou Dalagurr Food Sovereignty Lab & Traditional Ecological Knowledges Institute, the development of faculty book circles and professional development trainings on Indigenous knowledges and histories, partnerships with local nonprofit organizations to integrate Indigenous perspectives into local high school curricula, and ongoing community-facing events.

2. *History of Native California:* https://www.youtube.com/watch?v=T-azcPugmKQ&t

3. *TEK & PBLCs* https://www.youtube.com/watch?v=liKV74avPso

Scheduled for its grand opening in fall 2023, the Rou Dalagurr Food Sovereignty Lab & Traditional Ecological Knowledges Institute will be the first Food Sovereignty lab in the CSU system and will build a space that supports Tribal communities in ongoing revitalization of traditional ecological knowledges and Indigenous sciences. The development of Rou Dalagurr has been an interdisciplinary and collaborative effort that is student-designed and community-informed. This project was designed and proposed by students in fall 2019 as part of the NAS 331: Indigenous Natural Resource Management Practices course. These students represented twelve different majors from both the College of Arts, Humanities & Social Sciences and the College of Natural Resources & Sciences. Students worked collaboratively during the semester to interview Native faculty,staff and-Native students on campus, and also held a community stakeholders meeting to discuss the implementation of a Food Sovereignty Lab. This research was subsequently awarded second place in the Social Sciences Graduate Research division of the CSU Student Research Competition in April 2020.

The Rou Dalagurr Food Sovereignty Lab & Traditional Ecological Knowledges Institute sits adjacent to the Native Forum and Goudin'i Gallery outside the BSS building. In November 2022 the outdoor space was officially renamed "Wiyot Plaza" (Humboldt Now 2022). Wiyot Plaza includes an official space allocation to the NAS Department and "Shared Maintenance and Co-Management Agreement" developed with Facilities Management and approved on July 6, 2022—a historic and transformative formal exterior space allocation. The agreement 'inscribe(s) the University's shared commitments to caring for the space, and to relationship with the landscape and the more-than-human relatives therein, ensuring their well-being, to return tribal community access and sovereignty to landscape, and to ensure the safety of those who utilize it. It is the first landscape co-management agreement between an academic department and the University and required ongoing work by the Food Sovereignty Lab Research Associate, Karley Rojas, along with Co-Directors Dr. Cutcha Risling Baldy and Dr. Kaitlin Reed to amend, navigate, and negotiate campus policies. Rou Dalagurr is a Wiyot term that means 'everyone works/work together/everyone work.'[4] The governance of the lab is made up of a majority Native American leadership. The steering committee, which includes representatives from local Tribal Nations and Indigenous organizations/nonprofits, meets bi-monthly. The Steering Committee also has students and youth members. In 2022 the lab was awarded a $1 million dollar grant for the "Food for Indigenous Futures" project from the Sierra Health Foundation for implementation of youth food sovereignty camps; the lab also hosts the Indigenous Foods Festival at Cal Poly Humboldt; and is building ongoing workshops and interventions into curriculum that centers food sovereignty and TEK. Wiyot Tribal Chairman, Ted Hernandez, illustrates the potential significance of such a space on our campus:

> Such a lab could provide hands on experience for students and tribal members in fields such as botany, biology, genetics, anthropology, Native American Studies, and Natural Resources Management, to name a few. The end result would be an interdisciplinary learning lab worthy of Humboldt that would both attract Indigenous people and students from out of the area, while also serving the local Indigenous peoples by helping to preserve their food sovereignty and native food security (NAS n.d.).

NAS continues to be mindful of how TEK is presented, acknowledged, and integrated into ongoing curriculum, projects, and initiatives at our university. Historical exploitation of TEK and Indigenous knowledge means that Indigenous peoples and Indigenous disciplines and departments are necessarily cautious about how TEK is framed within training and outreach programs. As Jessica Hernandez explains "We cannot just solely identify Indigenous teachings and remove Indige-

4. Listen to a pronunciation by Marnie Atkins: https://drive.google.com/file/d/15zj9tXsVmWlWbpmvGkqLo-eHCeEsK-k6P/view

nous peoples out of this narrative as co-option of Indigenous knowledge contributes to the oppressive narrative we currently have in the environmental discourse. These teachings do not and cannot be applied in the Western scientific paradigms or frameworks without incorporating Indigenous peoples as well" (Hernandez 2022:108). In other words, TEK cannot happen without Indigenous peoples, and by extension, we cannot create a PolyTEK without Indigenous people, communities, and nations.

Beginning in spring 2021, the Cal Poly Humboldt Center for Teaching & Learning, the NAS Department, the Office of the Provost and the Office of Sustainability held a series of decolonial coffee breaks in the format of a book discussion group for faculty. The text for this book circle was *As Long As Grass Grows: The Indigenous Fight for Environmental Justice, from Colonization to Standing Rock* by Dina Gilio Whitaker (Colville Confederated Tribes). The specific objective for this program was to hold a book circle that would (1) facilitate evaluation of interest and needs/gaps for future programming on similar topics, and (2) if merited, serve as the first step in a multi-semester Faculty Learning Community on Indigenous environmental justice, TEK, sustainability discourse, and decolonization to facilitate productive further learning. All disciplines were invited to creatively engage with this challenge. Due to the success and positive feedback from faculty that participated in this initial book circle, this type of professional development activity continued into the following academic year. During Fall 2021, the faculty book circle read *We Are the Land: A History of Native California* by Damon B. Akins and William J. Bauer Jr. (Wailaki/Concow). During Spring 2022, the faculty book circle read *Braiding Sweetgrass: Indigenous Wisdom, Scientific Knowledge, and the Teachings of Plants* by Robin Wall Kimmerer (Potowatami).

In addition to the faculty book circle, the NAS Department and the Office of Sustainability collaborated to host the Decolonizing Sustainability Speaker Series entitled *Decolonizing Sustainability: Amplifying Indigenous Perspectives and Transforming Sustainability Discourse* during fall 2021. This speaker series highlighted and unpacked intersections of settler colonialism, white supremacy, and systems of power/privilege/oppression within the discourse and rhetoric of contemporary sustainability, environmental, and climate change movements. This series was intended to serve as a deep dive into the problematic and harmful discourse(s) of sustainability with the goal of moving toward a decolonial sustainability movement that amplifies Indigenous sovereignty and experience. Indigenous Science and NAS are still left out of much of the discourse around "sustainability" as TEK challenges the "hegemonic visions of sustainability discourse" (Velasco-Herrejón, Bauwens, and Friant 2022: 1). Velasco-Herrejón, Bauwens, and Friant elaborate that in sustainability discourse, "it is assumed that through these innovations, economic growth could be effectively "decoupled" from environmental degradation, leading to a dematerialized and decarbonized "green economy" created by a third or even fourth industrial revolution" (2). They conclude "different conceptualizations of sustainability must be recognized to ensure an inclusive and just energy transition" (1).

The Decolonizing Sustainability Speakers Series included talks entitled:

- We Have Always Been Scientists: Western Science, Sustainability & the Delegitimization of Indigenous Knowledge Systems;
- Nobody Asked You John Muir: Settler Colonial Environmentalism, White Supremacy & Sustainability;
- Colonialism Is Not Over: Green Colonialism & Sustainability from the TMT to the Terra Gen Wind Project; and
- Indigenous Knowledge Cannot Be Treated as a Back Up Plan: Indigenous Collaboration, Sustainability & Decolonization.

The sessions were recorded and are available to be accessed, used, and built into curriculum at the NAS YouTube page.

The NAS Department has also collaborated with community partners, including Save California Salmon, the Blue Lake Rancheria, the Yurok Tribe's Visitor Center, Klamath-Trinity Joint Unified School District's Indian Education Program, Humboldt County's Pathmakers Program, and the Hoopa High School Water Protector's Club to produce curricula that centers Indigenous perspectives and traditional ecological knowledge. In spring 2021 NAS collaborated with Save California Salmon on the Advocacy and Water Protection in Native California High School Curriculum and Teacher's Resource Guide. The curriculum, which meets California state standards in science, social studies, health, history and language arts, responds to California's urgent water, climate and educational crises, along with the need for Native American culturally informed education and representation in schools. The curriculum features online, classroom, and nature-based learning. As Dr. Risling Baldy notes:

> This curriculum was created for high schools, however all of California's primary and university students and faculty can benefit from the culturally-informed lessons it provides. California faces a water and climate crisis that will only be solved by foregrounding Indigenous management practices. Our restorative environmental management and tribal place-based knowledge are best practices for climate resiliency. If we teach the next generation how to better manage and live with the land. They will become the leaders that can solve our challenges. (2021).

Over the past twenty-five years, Cal Poly Humboldt has become a leader in both natural resources sciences and sustainability, as well as NAS. In the next 25 years, Cal Poly Humboldt is well poised to become a leader in Indigenous Environmental Studies & Sciences. While we have achieved significant milestones, there is certainly more to do. As part of implementing a PolyTEK institution, the university must develop additional programming and curricula that meets these needs. NAS has proposed and the Board of Trustees has approved the creation of a new 'Applied Science" Bachelor's of Science degree in Indigenous Science and the Environment to be housed in the NAS Department. This innovative major brings together principles of both Indigenous Knowledge and western science through curriculum and hands-on applied science research methodologies with a critical and integrative approach to complex environmental issues. Additionally, and significantly, Cal Poly Humboldt's designation as a Hispanic-Serving Institution offers a unique opportunity to build decolonial pedagogies and cultural affirming curriculum across Indigenous nations throughout the Western Hemisphere. To meet these curricular needs NAS was able to hire an additional faculty member, as part of AB1460 California Ethnic Studies funding and welcomed Dr. Cinthya Ammerman who focuses on global Indigenous studies, hemispheric relationality, land defense movements, and Indigenous climate change studies. Her current research follows the stories of various plants and the links they have created between Mapuche and California Native homelands mapping potential paths to hemispheric collaboration in response to climate change. Dr. Ammerman is a multi-heritage interdisciplinary scholar from Wallmapu, ancestral Mapuche homelands in southern Chile.

NAS is also a partner in the Masters of Engineering in Community Practice which will host the first cohort in fall 2023. "The one-year M.S. in Engineering and Community Practice degree addresses state and national workforce needs and is tailored to those who wish to pursue resource management positions that require strong technological and management skills with a particular focus on interfacing and working with Indigenous and traditionally underserved communities" (Engineering &Community Practice n.d.). Newly hired faculty Dr. Aaron Gregory (NAS) will work alongside Dr. Qualla Ketchum (Environmental Resources Engineering) to implement the program. Students will take both Engineering and NAS courses to help them navigate community-based research projects. Dr. Gregory's work is focused in science and technology studies, critical infrastructure studies and political ecology.

The results of continuing to expand our university engagement with TEK has also resulted in the hiring of two additional Native American professors in the College of Natural Resources and Sciences. Dr. Qualla Ketchum (Environmental Resources Engineering) and Dr. Daniel Lipe (Environmental Science & Management) are new additions to Cal Poly Humboldt as of 2022-23.

Toward a PolyTEK for All

There is a significant demonstrated need for Indigenous informed curricula, programming, and research. At the time of article preparation, there were roughly 190 job postings for "Tribal Liaison" or "Tribal Outreach" in federal government agencies, many of which focus on natural resources management or natural sciences. Many federal and state agencies have departments dedicated to Tribal relations and collaboration. Many non-profit environmental organizations and research institutions are also making commitments to improve their relationships and collaboration with Tribal Nations. There were over 20 positions at Tribal offices listed in Northern California specifically with a focus on both Indigenous governance and science. Tribal Nations are at the forefront of many ecological restoration efforts in this region and beyond. We anticipate that with the growth of Indigenous Science based research studies over the next decade, there will be an influx of students interested in the natural sciences that integrate TEK in a meaningful and hands-on way.

While we have made important strides at Cal Poly Humboldt, it is important to be cognizant of the power dynamics that exist between Indigenous knowledges and western institutions of higher learning, and the ways in which settler colonial ideologies continue to inform understandings of Native peoples, land, and environment. Historically, research in Native American communities has been very extractive, so we need to be particularly aware of developing equitable relationships with Tribal Nations and peoples who are partners and collaborators—not subjects from which the university can acquire knowledge. And thus, a polyTEK institution that foregrounds TEK must also develop and sustain collaborative policies. There are concerns about situations wherein TEK is marketed by the university, but access to university resources, support for higher education attainment, and support for Tribal peoples is still limited. These limitations continues an extractive relationship wherein the university profits off of Indigenous lands and knowledges—at the expense of Indigenous peoples. NAS is the discipline that was advocated for by Tribal nations. We have not been in the academy as a discipline as long as other disciplines have and yet, now that we see how NAS curriculum is important to future careers and the future of the planet, we also see an attempt to capitalize on NAS focused curriculum in other departments without acknowledging, collaborating, or partnering with NAS. Alongside our work to strengthen the Department, our faculty and staff must consistently prepare for challenges to our very existence as a department and major; respond to accusations of favoritism by administrators; and respond to administrative decisions that deprioritize NAS needs. The development of programs, projects, majors/minors, certificates and other areas of academic development and community facing projects should invite and include NAS as partners and encourage the support and involvement of a discipline that represents the vision of Tribal Nations and peoples and has historically been devalued by western academic disciplines and institutions. There are many examples we could offer to illustrate ongoing challenges but one clear example is the Food Sovereignty Lab. In 2019 students worked with Dr. Risling Baldy to initiate a space request for the lab. This initial request was denied by the University Space Committee. The committee noted in their denial of the lab that they wanted to reserve the space for a program that would serve the "general" student population. The implication was that a lab space for NAS would not be for "all students." In a response memo to the Space Committee the NAS Department wrote:

> Our major, minors, programs and projects are very popular across campus and with the local communities because Indigenous knowledge is foundational to how we will approach major issues like climate change. Dismissing Native knowledges as being too specialized or not for the "general student population" effectively stereotypes and propagates attitudes that have always functioned to marginalize and dismiss Indigenous ways of knowing, our philosophies, and our place in higher education. We are now being further marginalized as a department on this campus, by segmenting an NAS Center/Lab as not being for all students or stating that only a "general student space" will be accessed by all, as if we do not serve the general student population through our work, outreach, curriculum, and community partnerships. We wonder if these types of assumptions about whether or not our Lab/Center would serve a "general

student" population would be expected of other labs on campus and why NAS is seemingly held to a higher standard for being able to manage space as part of our department curriculum and program.

NAS was finally able to secure the space when the students appealed to the Academic Senate and provided public testimony about the importance of the lab. At that time the Space Committee reversed its decision but said it would only approve the space request "conditionally" as they required that NAS fully raise the funds for remodel before getting the space. It was clear from this experience that the NAS Department continued to be held to higher standards than other programs requesting facilities across campus. While we consistently met all of the requirements and now are in the process of remodeling the lab, this experience is just one way to highlight the ongoing challenges faced by the Department in just the past five years.

At the core of Cal Poly Humboldt is a commitment to social and environmental justice that is shaped by ongoing support for Indigenous peoples that moves beyond acknowledgement and toward an ethical praxis of being in place. Significantly, Humboldt's designation as a Hispanic Serving Institution offers a unique opportunity to build decolonial pedagogies and cultural affirming curriculum. This cannot be accomplished merely through the presence of Indigenous peoples on campus but must also include meaningful efforts to show how NAS is integral to the fundamental curriculum and learning outcomes of the institution. If Cal Poly Humboldt is truly committed to guiding students in "social and environmental sustainability" so they can make a difference in the world, we must continue to support the key role that Native American Studies plays in grounding students in decolonial frameworks and land-based learning across colleges.

When visiting Humboldt virtually for a talk to our campus regarding the role of TEK in the forthcoming Polytechnic brand identity and transformation, Dr. Kimmerer's response demonstrates her expert support for this ongoing work, a polyTEK that moves disciplines and education forward.

> Oftentimes, there's this notion that somebody else is going to talk about these things, but they don't really belong in the physics classroom, or they don't belong in the botany classroom. Well, they belong in every classroom, I think. And, giving people the tools to enact that. I think of how many colleagues I have who say, 'Well you know, I'd like to think about those things. I'd like to introduce them into my classroom, but I don't know how. I'm afraid of doing it wrong, so I don't do it at all.' But, I think [NAS/Faculty Book Circle] approach of faculty and staff professional development in this arena is the implementation step for this first priority of telling the truth. ...I want to sincerely reciprocate that gratitude, knowing the path that you all are on, and what your aspirations are to center Indigenous ways of knowing in your university is deeply inspiring to me, and I want to continue to learn from you.

Kimmerer's contributions here should not be overlooked because they underscore that Cal Poly Humboldt should continue investing in the work of faculty development to integrate TEK and Indigenous wisdom to the identity of the Institution. As a leading educator, best-selling author, and thought leader, Kimmerer's comments demonstrate that the campus is already pushing the boundaries of innovation by moving the needle in the way we teach to a lens that foregrounds history and Indigenous knowledge *and* that we should continue doing so. These comments align with all the positive feedback from faculty and their demand for future programs to go deeper with this work.

TEK must be ethically practiced and culturally informed. In short, TEK is not a singular module or overview taught in a class, it cannot be treated as a diversity initiative or as window dressing for primarily western scientific courses, majors or programs. There needs to be meaningful support from higher education institutions like Cal Poly Humboldt for Indigenous led movements like #LandBack and the return of Indigenous remains and cultural items from university collections. As a campus that trains thousands of students on the best practices of environmental management, it is imperative that Humboldt teach in a way that honors, respects, and elevates the need for Indigenous sovereignty in environmental and natural resource work.

Resources

Akins, Damon B., and William J. Bauer Jr. 2021. *We Are the Land: A History of Native California. Oakland,* CA: University of California Press.

Anderson, M. Kat. 2005. T*ending the Wild: Native American Knowledge and the Management of California's Natural Resource*s. Berkeley: University of California Press.

Blackburn, Thomas C., and Kat Anderson, eds. 1993. *Before the Wilderness: Environmental Management by Native Californians* (Vol. 40). Menlo Park, CA: Ballena Press.

Cook-Lynn, Elizabeth. 1997. Who stole Native American Studies? *Wicazo Sa Review* 12(1):9-28.

Council of American Indian Faculty and Staff (CAIFS). 2021. Statement on PolyTech University Development and Indigenous Knowledge/Indigenous Science/ Traditional Ecological Knowledge. Available as appendix.

Cronon, William. 1996. The trouble with wilderness: or, getting back to the wrong nature. *Environmental History* 1(1):7-28.

Deloria Jr., Vine. 1989. American Indians and Humboldt State. Retrieved April 18. 2023. http://hdl.handle.net/2148/1244.

Engineering & Community Practice. N.D. Majors and Programs. Retrieved April 18, 2023. https://www.humboldt.edu/programs/engineering-community-practice.

Gilio-Whitaker, Dina. 2019. *As Long as Grass Grows: The Indigenous Fight for Environmental Justice, From Colonization to Standing Rock.* Boston, MA: Beacon Press.

Hernandez, Jessica. 2022. F*resh Banana Leaves: Healing Indigenous Landscapes through Indigenous Science.* Huichin (unceded Ohlone land aka Berkeley, CA): North Atlantic Books.

Humboldt Now. 2022. 'Wiyot Plaza' is new name for campus space. Retrieved April 18, 2023. https://now.humboldt.edu/news/wiyot-plaza-new-name-campus-space.

Humboldt State University (HSU). 2021. Polytechnic Prospectus. Retrieved April 18, 2023. https://www.humboldt.edu/sites/default/files/hsupolytechnic-prospectus.pdf.

Johnson, Matthew D., Amy E. Sprowles, Katlin R. Goldenberg,Steven T. Margell, and Lisa Castellino. 2020. Effect of a place-based learning community on belonging, persistence, and equity gaps for first-year STEM students. *Innovative Higher Education* 45:509-531.

Kidwell, Clara S. 2005. *Native American Studie*s. Lincoln, NE: University of Nebraska Press.

Kimmerer, Robin Wall. 2015. *Braiding Sweetgrass: Indigenous Wisdom, Scientific Knowledge and the Teachings of Plants.* Minneapolis, MN: Milkweed Editions.

LaDuke, Winona. 2019. "Foreword: In Praise of Seeds and Hope." Pp. xiii–xvi in *Indigenous Food Sovereignty in the United States: Restoring Cultural Knowledge, Protecting Environments, and Regaining Health.* Norman, OK: University of Oklahoma Press.

Native American Studies (NAS). N.D. Rou Dalagurr Food Sovereignty Lab and Traditional Ecological Knowledges Institute. Retrieved April 18, 2023. https://nasp.humboldt.edu/fsl.

Norgaard, Kari M. 2019. *Salmon & Acorns Feed Our People: Colonialism, Nature & Social Action.* New Brunswick, NJ: Rutgers University Press.

Pierotti, Raymond, and Daniel Wildcat. 2000. Traditional ecological knowledge: the third alternative (commentary). *Ecological Applications* 10(5):1333-1340.

Ramos, Seafha. 2019. "Sustaining Hlkelonah Ue Meygeytohl in an Ever-Changing World." Pp. 85–93 in *Ka'm-t'em: A Journey Toward Healing.* Pechanga, CA: Great Oak Press.

Reed, Kaitlin. 2023. *Settler Cannabis: From Gold Rush to Green Rush in Indigenous Northern California.* Seattle, WA: University of Washington Press.

Risling Baldy, Cutcha. 2021. "Advocacy and Water Protection in Native California Curriculum." Save California Salmon. Retrieved April 24, 2023. https://www.californiasalmon.org/

Spence, Mark D. 1999. *Dispossessing the Wilderness: Indian Removal and the Making of the National Parks.* Oxford, UK: Oxford University Press.

Velasco-Herrejón, Paola, Bauwens, Thomas, and Friant, Martin Calisto (2022). "Challenging dominant sustainability worldviews on the energy transition: Lessons from Indigenous communities in Mexico and a plea for pluriversal technologies." *World Development* 150:105725.

Voyles, Traci B. 2015. *Wastelanding: Legacies of Uranium Mining in Navajo Country.* Minneapolis, MN: University of Minnesota Press.

Wildcat, Daniel R. 2010. *Red alert!: Saving the Planet with Indigenous Knowledge.* Wheat Ridge, CO: Fulcrum Publishing.

White House Office of Science & Technology Policy & Council on Environmental Quality. (2021). *White House Commits to Elevating Indigenous Knowledge in Federal Policy Decisions* [Press Release] .Retrieved April 18, 2023. https://www.whitehouse.gov/ostp/news-updates/2021/11/15/white-house-commits-to-elevating-indigenous-knowledge-in-federal-policy-decisions/.

Appendix:

Council of American Indian Faculty and Staff (CAIFS)
Humboldt State University
Statement on PolyTech University Development and Indigenous Knowledge/Indigenous Science/ Traditional Ecological Knowledge
April 29, 2021

The Humboldt Council of American Indian Faculty and Staff (CAIFS) is an advisory council that includes members from many of our Humboldt Departments, Programs and Colleges across campus. This multi-tribal, interdisciplinary group meets regularly to discuss complex issues facing our American Indian communities with a focus on campus and regional matters. One of our objectives is to provide feedback on issues as a collective of tribal voices with the end goal of ensuring that we are advocating for and supporting a campus that meets the needs of our tribal students and tribal community while also building a positive campus environment for all of our campus and local community.

campus and local community.

With Humboldt undertaking a self-study to explore becoming a polytechnic university, CAIFS has developed this guiding document to help ensure that tribal voices are not only included but become a part of the foundational fabric and ongoing leadership of this opportunity in front of us. Within the stated goals of this study, we have included guidance to build on the concepts of sustainability and hands-on learning, as well as infusing our polytechnic goals with Traditional Ecological Knowledge, and equitable and ethical practices. Each of these speaks directly to the strengths and wisdom that American Indian and Indigenous communities and peoples can offer in this process. We are fortunate to have as a part of CAIFS a number of cultural practitioners, community advocates, community organizers, and community leaders with the experience and knowledge to build best practices for engaging Indigenous science and TEK in a meaningful and informed way.

We believe in working directly with tribal communities. Humboldt is positioned to offer a cutting edge and unrivaled polytechnic experience to current and future students. Humboldt has the opportunity to build on all of these strengths and we believe this work must be undertaken with intentionality and integrity so that we can take that step forward together. Indigenous knowledge systems are especially important and appropriate to consider in the development of a polytechnic institute because Indigenous knowledges are fundamentally **interdisciplinary** and **applied**. Indigenous knowledges are also at the forefront of cutting-edge research interventions in the sciences and western academic institutions. When we talk about or propose "decolonizing" curriculum or higher education we must build this from Indigenous frameworks with Indigenous Peoples at the center of our academic vision and planning.

What is Indigenous Knowledge (IK)?

Indigenous Peoples compose 6-8% of the population globally (approx. 350 million) and 1.5% of the United States population (approx. 4.1 million peoples). All Indigenous groups come from distinct lands, cultures, languages, worldview, philosophies, and ways of knowing. Indigenous Peoples have millennia-old Indigenous Knowledge (IK) systems that are tribally and geographically specific. Indigenous Knowledge is also referred to as traditional ecological knowledge (TEK), tribal knowledge, tribal science, Native science, Indigenous environmental science, Indigenous environmental studies.

Indigenous knowledge systems are diverse and they are rooted within specific cultural and geographical contexts. However, there are important distinctions between Indigenous bodies of knowledge and Western/colonial knowledge.

1. IK is communal, not individual. Some IK includes culturally-sensitive information that tribal nations may not choose to share with researchers or universities. Some IK includes information that can, and should, be accessed by all, including Indigenous perspectives on law, business, government, technology, health, art, history, etc.
2. It is embedded in community practices, rituals, relationships and is difficult to codify.
3. Valid in its own right and does not need to be verified or legitimized by other bodies of knowledge.
4. IK is not frozen in time; some knowledge adapts to reflect the dramatic changes reoccurring within Indigenous communities today.
5. Tribal peoples are sovereign nations and have the right to exercise self-determination over their knowledge systems—therefore, Humboldt needs to work in partnership with tribal nations—not extract knowledge from.
6. Lastly, IK is fundamentally interdisciplinary.

Within the context of California, Indigenous peoples have lived with and stewarded their lands in this region from time immemorial. Native peoples in California developed sophisticated and complex ecological management regimes[a] that promoted habitat heterogeneity and increased biodiversity.[b] The invasion of northern California by settlers facilitated land theft and genocide of Native peoples—but also the erasure of land management practices (e.g., creating "wilderness") and criminalization of Indigenous land management practices (e.g., fire suppression policies). The massive ecological transformation in northern California—brought on by 170 years of settler colonial invasion—cannot be understood without an understanding of settler colonialism and will require Indigenous knowledge and practices to rectify.

Indigenous Knowledge systems pre-date settler colonial invasion and the establishment of Western universities. To conduct ethical research and facilitate reciprocal collaborations with tribal partners, requires an understanding of the historical context of the Western university and its role in settler colonialism. Western academic disciplines are all implicated in imperialism and colonialism.[c] Devon Mihesuah and Angela Wilson note that "The academy has much invested in maintaining control over who defines knowledge, who has access to knowledge, and who produces knowledge."[d] The creation of a polytechnic institution must actively challenge historic patterns of domination through the prioritization of Indigenous knowledge systems and values in its creation and implementation process. This requires developing and maintaining ethical relationships with tribal nations. In nurturing these relationships, we must always have the best interests of the native community at heart. While we are a place of education, we should first be advocates and protectors of Indigenous culture and knowledge. We must make a distinction in our decision-making between knowledge that has a place in the university and knowledge that belongs in the community. We must be cognizant of how treasured knowledge of our Indigenous peoples may be exploited and consumed. In the development of the polytechnic self-study, we offer the following guidelines and recommendations for best practices in engaging Indigenous communities and foregrounding Indigenous knowledges.

Critical Approached to Sustainability and Climate Resiliency

Part of the intervention that is made by Indigenous scholars and Indigenous community knowledge holders is to offer critical analysis of sustainability, conservation, and other western environmental perspectives that can be developed in a way that continues the same exploitative practices that have caused the problems we face. We caution against superficial comparisons between Western notions of sustainability and the ethos of living well in Indigenous knowledge systems. Deborah McGregor (2004) argues that "Indigenous views of sustainable development are concerned with giving rather than taking, and with what it is that we can contribute to creation. Indigenous views also include active resistance (sometimes to sustainable development itself) and the process of reclaiming our traditions" (77). Rhetoric of sustainable development has been used to continue the process of Indigenous land dispossession. This is often referred to as "green colonialism." According to a World Bank report of 2008 Indigenous peoples make up 5% of the global population and yet, protect 80% of the globe's biodiversity. Green industries and conservation entities target Indigenous territories, using the lack of ecological footprint as rationale for seizure and exploitation under the guise of "sustainability" and "sustainable development." Discussions at Humboldt and as part of building a Polytechnic university should include Indigenous peoples and curriculum that foregrounds a critical approach to sustainable development.

Anthropogenic climate change is an intensification of environmental change imposed on Indigenous peoples by colonialism and capitalism.[5] The field of Indigenous Climate Change Studies has emerged to support Indigenous peoples' and is reflected in Indigenous knowledge systems and relationships to land, as well as climate resilience plans created by tribal nations. Indigenous Climate Change Studies emerges from the memories, knowledges, histories, and experiences of oppression of Indigenous peoples that differ from many of the non-Indigenous scientists, environmentalists, and politicians that are prominent in the framing of the issue of climate change today. Responding to colonialism-induced climate change requires a return to traditional ecological knowledge (e.g., burning the landscape). We encourage critical reflections on curriculum, programs, and research that is proposed or recommended for the Polytechnic self-study to foreground how to best include an Indigenized curriculum.

Guidelines for TEK and Partnership with Indigenous Communities

1. For far too long research has been utilized to disempower Indigenous peoples and communities. It is important that named partnerships and opportunities for engaging TEK and Indigenous science are done in a way that is informed by Indigenous partnerships and relationships. TEK must be ethically practiced and culturally informed. There are many considerations when engaging with TEK especially around sustainable use, and it is also important that engagement with TEK and Indigenous science not only center knowledge sharing, but also how departments, programs, and colleges are dedicated to upholding sovereignty and self-determination and working to empower Indigenous students, communities, and ongoing projects of land return, environmental justice, and education.
2. Tribal nations should be considered equal partners in projects that are being developed on their lands or in their tribal communities. Consultation is not collaboration. A polytechnic that foregrounds TEK should also be clear to develop and sustain collaborative policies.
3. Tribal programs and projects hosted at Humboldt should be prioritized and there should be support for these programs across campus. We have a number of leading programs through Indigenous Peoples Week (October) and California Indian Big Time (April) that would benefit from institutional support.
4. To be a leading polytechnic means naming and engaging with TEK across this campus and it is important that courses that center Indigenous perspectives, (such as Native American Studies, American Indian Education, Decolonizing Social Work, etc.) not be treated as an "elective" to majors but instead as essential courses to our campus curriculum. Indigenous knowledge systems are relevant to every field on campus, ranging from health, art, business, law, governance, sciences, etc. We have some outstanding examples and current practices to build

upon, including both the ongoing work to introduce a NAS focused Forestry concentration as well as with our cross-disciplinary work in the Place Based Learning Communities (PBLC's). NAS is an important and influential component of the PBLC's at Humboldt and beginning in Fall 2020, Humboldt is the only CSU where a majority of the incoming first-year students who are majoring in the sciences will also have taken a NAS course. A survey of Klamath Connection students, 87% felt that the knowledge they gained about Native American cultures will help them in their careers. Qualitative data also indicates that the NAS course component of the PBLCs is very influential in students' decisions to remain at Humboldt. There are still majors on this campus that do not include Indigenous perspectives as integral to their curriculum and we see value in offering support for curriculum revision that truly embraces Indigenous knowledges as part of all majors across campus.

Recommendations for Polytechnic Self-Study/Indigenous Campus

- **Support the creation of a Vice-President of Tribal Affairs office** to increase communication, empower CAIFS and help faculty and staff understand best practices for collaboration. Many other campuses in the CSU have a Tribal Liaison position and as we increase our reliance on Indigenous communities and tribal nations for our curriculum and programs, we also need to provide them with representation moving forward. Collaborating with CAIFS to help design and implement this position will be important.
- **Provide additional support for Indigenous students:** Embracing and utilizing TEK as central to our Polytech curriculum also necessitates a dedication to our Indigenous students so they can engage in programming that supports their student success, and offering opportunities for Indigenous students to attend and succeed at Humboldt (through fee/tuition waivers, scholarships, research assistantships etc.) demonstrates how our university will give back to our communities.
- **Support the continued enrollment and growth of Indigenous students and Indigenous student focused programming on campus:** It is essential to establish a polytechnic university that embraces BIPOC students and we encourage finding ways to evaluate student admissions based on a holistic review. We urge the university to dedicate additional resources to Native American student recruitment and encourage the creation of a Native American Student Recruiter position. We also recommend providing additional resources to our Native programs like ITEPP and INRSEP+ to outreach, review, and guide Native students through the application and admissions process.
- **Provide support for the Native American Studies Department** to grow their tenure-track faculty, course-offerings, curriculum development and continued outreach.
- **Provide support for active leadership of CAIFS** at our university including funding of administrative and faculty positions to assist with the ongoing development of recommendations for tribal engagement.
- **Reconstitute the Center for Indian Community Development (CICD):** CICD was a center for language and outreach to tribal peoples and tribal nations. This community-facing program was one place where Native peoples could build partnerships with the University but it was also a community focused program at Humboldt that demonstrated how Humboldt was dedicated to tribal community success. Programs like this build community support for campus work and if we continue to rely on tribal leaders and community members, we should also provide resources for the community that is focused on hands-on applied assistance.
- **Develop incentives for a cluster-hire of Native faculty across campus** with a focus on the College of Natural Resources and Sciences (CNRS). While there have been qualified Indigenous candidates with connections to local community, there are currently no tenure-track Native faculty in CNRS, especially the STEM programs. There are several examples across multiple campuses throughout the

country who have been able to integrate TEK/Indigenous Science into faculty hires in the STEM disciplines. Efforts must be made, not only to hire Native faculty across campus, but convert non-tenure track faculty to tenure track positions.

- **Provide ongoing funding for Native focused community and student centered programs**. We see a great need for ongoing increased funding for native specific spaces on campus that allow students, faculty, staff, and community to encounter diverse perspectives and provide unique places to experience Indigenous knowledges that will enhance the overall polytech experience. There are so many important programs on this campus that will need additional support: the NAS Food Sovereignty Lab, ITEPP, INRSEP, American Indian Education Minor, and the Social Work Program. We see great value in putting together a working committee to help design an Indigenous space plan for the campus that can be included as part of the overall space planning goals for the University.

We look forward to how we can continue to engage the Polytechnic Self-Study in a conversation about the role of Indigenous peoples and nations in the future of our university. We encourage the leaders of the self-study to attend a CAIFS meeting to gather additional insight and recommendations. We also offer some potential program links of interest as potential best practices or directions for moving this study forward.

1. Indigenous Strategic Plan (University of British Columbia): https://indigenous.ubc.ca/indigenous-engagement/indigenous-strategic-plan/
2. NAGPRA and applied sciences https://www.nps.gov/orgs/1539/nas-nagpra-and-applied-science.htm
3. Public Health Training Certificate in American Indian Health https://caih.jhu.edu/training/scholars/category/public-health-training-certificate-in-american-indian-health
4. Applied Science in Legal Studies (with a focus on Federal Indian Law) https://catalog.uaa.alaska.edu/undergraduateprograms/coh/legalstudies/aas-paralegalstudies/
5. Applied Indigenous Studies: Bachelor of Science (Northern Arizona University) https://nau.edu/ais/bs/#:~:text=Applied%20Indigenous%20Studies%2C%20Bachelor%20of,communities%20into%20the%2021st%20century.
6. Indigenous Environmental Studies & Sciences, B.A. or B.S. (Trent University) https://www.trentu.ca/iess/

HJSR
50TH ANNIVERSARY EDITION

Fostering Socio-Ecological Resilience to Wildfire by Interconnecting Knowledge Systems at Cal Poly Humboldt

Jeffrey M. Kane *<jk55@humboldt.edu> (Cal Poly Humboldt)*, **Erin C. Kelly** *(Cal Poly Humboldt)*, **Benjamin C. Graham** *(Cal Poly Humboldt)*, **David F. Greene** *(Cal Poly Humboldt)*

Abstract

The wildfire-related challenges of Northern California and many other regions in the western United States are daunting in scope and magnitude. Ecologically and culturally salient solutions that limit the negative impacts of wildfire and promote resilience of human and ecological systems will require newer approaches. Through Cal Poly Humboldt and the Fire Resilience Institute, there is greater emphasis on the interconnection of knowledge systems across education, training, research, and management. Here we highlight several on-going efforts that seek to enhance the fire resilience workforce, promote socio-ecological resilience through interdisciplinary projects, and inform management through monitoring and research projects that intentionally incorporate multiple knowledge systems. Shifting to a more inclusive process has many potential benefits but will also pose challenges and require modification of approaches. Here we emphasize some on-going efforts at Cal Poly Humboldt to intentionally bridge knowledge systems to make advances on wildfire-related challenges. Socio-ecological resilience and coexistence with fire can be fostered but the long-term effectiveness will greatly benefit from approaches that are inclusive, equitable, and interconnected across the many stakeholders affected and disciplines involved.

Keywords: co-production, fuels management, interdisciplinary, knowledge exchange, prescribed fire, wildfire

Fire is an ecological and cultural process that has long contributed to the biodiversity of California (van Wagtendonk et al. 2018). Over the last several decades, the drying of fuels induced by rising temperature has been one of the main causes of the increased frequency, size, and severity of fires in the western US (Abatzoglou and Williams 2016; Westerling 2016; Parks and Abatzaglou 2020). The other major contributor to these trends is forest densification that resulted from past state and federal land management policies (fire suppression and intensive logging), and the removal of Indigenous burning practices. California now faces burgeoning costs and wide-scale adverse impacts to human communities and ecosystems (Steel et al. 2015, Syphard et al. 2017, Mann et al. 2016, Hoover 2020).

Much of the forestland in the Western United States is managed by federal agencies. Until recently, fire suppression has remained the de facto national management approach despite a growing body of knowledge emphasizing the ecological and social harm of attempting to stop all fire (Quigley et al. 1996; Stephens and Ruth 2005). Butler and Goldstein (2010) claim that land management institutions like the USDA Forest Service are "caught in the rigidity trap" of fire suppression, which is reinforced through "incentive structures, agency budgets, and professional practice" (1), as well as by the encouragement of many rural citizens. The ongoing crises of large-scale wildfires can reinforce financial and/or political support for the status quo as people implement previous strategies on which they have been trained (Yaffee 1996; Stephens and Ruth 2005; McCaffrey 2015). Despite evidence of maladaptation, including a reduction of productivity in ecosystems and damage to property and lives, rigidity traps are persistent because of risk aversion and learned behavior among stakeholders (Gunderson and Holling 2002; Butler and Goldstein 2010; Nair and Howlett 2016; Cumming 2018). In a rigidity trap, new information or circumstances may not result in changed behavior, as "there is no explicit need to learn, because the bureaucracy already knows what should be done" (Lebel et al. 2011: 50).

At the federal level, there is evidence of a culture shift in land management agencies such as the Departments of Agriculture and the Interior. They have created innovative policies to address threats to human communities and residences from changed wildfire patterns. These policies call for large-scale, cross-ownership boundary fire and fuels management projects that proactively address wildfire risk and re-incorporate fire as a beneficial ecological and cultural process (USDA Forest Service 2012; North, Collins, and Stephens 2012; Steelman 2016; Bixler et al. 2016). Policies include the 2014 National Cohesive Wildland Fire Management Strategy[1] to create resilient landscapes and protect homes and communities, and funding initiatives that incentivize large-scale and multi-landowner fuels treatments, including the Collaborative Forest Landscape Restoration Program, and USDA and USDOI Joint Chiefs funding (Butler and Schultz 2019, Schultz, McCaffrey, and Huber-Stearns 2019; Kelly, Charnley, and Pixley 2019). At non-federal levels experimentation and strong local leadership (often among municipal or county governments, Tribes, or non-governmental organizations) have helped to re-introduce a fire culture built on resilience rather than reaction. Much of this work is grounded in the efforts of Tribal land managers. Long and Lake (2018) trace the displacement of Tribes in the Western U.S. as a major part of the wildfire suppression rigidity trap and note Tribal re-engagement in landscape management as a solution.

The problems associated with wildfire in California are multifaceted and complex with impacts that are increasingly affecting a greater number of human communities and ecosystems. Socio-ecological resilience involves the capacity of systems to endure, adapt, and influence change in the presence of disruptions and to innovate and transform into new, more desirable configurations in response to disturbance (Folke 2006). This perspective highlights the need to think of human communities and ecosystems as coupled systems whose interconnections are essential for fostering resilience (Moritz et al. 2014). Enhancing socio-ecological resilience to wildfires will benefit from new formulations of concepts and approaches in how communities produce, exchange, and use knowledge. Some existing knowledge systems related to fire management include traditional ecological knowledge (also referred to as Indigenous knowledge), western scientific knowledge, and local empirical knowledge (managers and community members). In many cases, knowledge of fire is often underutilized, and fire management decision-making may not integrate relevant skills and approaches across multiple sources (Hunter 2016, Adams et al. 2017). While advances have been made in creating better connectivity among knowledge systems through greater emphasis on management-relevant research and monitoring (LeQuire 2011, Shuman et al. 2022), co-production of knowledge (Norström et al. 2020), and knowledge exchange (e.g., Fire Science Exchanges, Maletsky et al. 2018), barriers and opportunities persist (Kocher et al. 2012, Hunter, Collavito, and Wright 2020, Glenn et al. 2022).

Sound approaches to socio-ecological resilience with respect to fire-related challenges will require better involvement of Indigenous communities to lead and share in decision-making processes. It is particularly

1. The National Cohesive Wildland Fire Management Strategy and updates are available at https://www.forestsandrangelands.gov/strategy/.

vital to respect and elevate knowledge of Indigenous communities in fire research and teaching. There are historical and current reasons that partnerships between non-Indigenous and Indigenous groups have been problematic because of misalignment of knowledge systems, governance, and institutions. As an example, Indigenous firefighters in California and Australia reported barriers and resistance to Indigenous concerns in implementing firefighting measures (Eriksen and Hankins 2014), even though Indigenous people in both locations have long-standing knowledge of the impacts and importance of fire in their landscapes. But institutional inertia and biases restrict the incorporation of indigeneity into fire management. Similarly, incorporating Indigenous knowledge and questions into research requires that we be willing to make significant changes to how we conduct our work—including sustained engagement with Tribes over the long term, receptiveness to new knowledge, and a willingness to be adaptive. These are attributes that are not always supported in typical grant-funded research, which comes with strict deadlines and deliverables.

In Northern California, the transition of Humboldt State University to California State Polytechnic University, Humboldt and the creation of the new Fire Resilience Institute create a foundation to address wildfire-related challenges. By expanding engagement in interdisciplinary programs and projects, these changes more holistically combine socio-ecological perspectives and foster greater interconnection of knowledge systems including Indigenous knowledge. In 2021, an interdisciplinary group of Northwestern California researchers and regional partners engaged with fire research, education, training, and management to form the Fire Resilience Institute. The overarching purpose of the Fire Resilience Institute is to serve as a center of expertise in fire science, ecology, and management that cultivates partnerships and fosters interdisciplinary research, education, training, and outreach to promote effective solutions to wildfire-related challenges while incorporating the need for, and importance of, fire as an ecological and cultural process. Starting in 2022, Cal Poly Humboldt became the third polytechnic university in California. Leveraging existing strengths in science, technology, engineering, and mathematics, including the largest forestry and fire management program in the state, and with an infusion of financial support from the state, the university is developing new programs that seek solutions to these 21st century challenges.

The 2023 launch of a Cal Poly Humboldt Bachelor of Science in Applied Fire Science and Management provides students with an interdisciplinary education focused on developing practical knowledge and skills to become fire science or fire management professionals. The wildfire-related challenges of Northern California and beyond are daunting in scope and magnitude. However, this set of challenges provides a unique opportunity following the transition to Cal Poly Humboldt and the creation of the new Fire Resilience Institute to facilitate the production and exchange of knowledge, employ new approaches to enhance engagement across stakeholder communities, and implement ecologically and culturally sound solutions to promote coexistence between people and fire.

The main objectives of this article are to describe the ways that Cal Poly Humboldt and the Fire Resilience Institute are directly addressing some of the grand wildfire-related challenges, with specific emphasis on: 1) enhancing the fire resilience workforce; 2) promoting socio-ecological resilience to wildfire; and 3) interconnecting fire knowledge systems to foster resilience under a new fire future. While these approaches have the capacity to make substantive contributions, it will require continued reflection and adaptation to promote and maintain advances.

Enhancing the Fire Resilience Workforce

The current wildfire crisis in California has prompted the USDA Forest Service (USFS) and the State of California to set an ambitious commitment to collectively treat 1,000,000 acres per year by 2025 (USDA Forest Service and State of California 2020). While these targets are laudable, attainment requires the rapid expansion of the fire and fuels management workforce in the state. The existing pool of trained and experienced professionals are already over-extended in many regions. Thus, there is a strong need to rapidly educate and train new professionals who can quickly contribute to fuel reduction efforts across many different ownership types. However, successful implementation of fuel reduction treatments also requires practitioners have a balance of education, training, and experience to increase the pace and scale of effective treatments that are ecologically and culturally sound.

Efforts to reincorporate fire as a management tool have expanded across the Western US. The call to increase prescribed fire to a "landscape scale" has become commonplace in the recent scientific literature (Noss et al. 2006; North et al. 2015) and in policy (Wurtzebach and Schultz 2016). However, efforts to scale up prescribed fire face many challenges, including legal and policy barriers, institutional inertia, and negative public perception (Quinn-Davidson and Varner 2012; Ryan, Knapp, and Varner 2013; Melvin 2018). Though legal impediments are frequently cited as constraints, recent research indicates that the most notable barriers may be related to capacity, including availability of qualified people, and the incentives to conduct burning (Schultz, McCaffrey, and Huber-Stearns 2019). This perspective suggests the need to not only train people in prescribed burning, but to also nurture a culture of burning, forest restoration treatments, and to build upon partnerships in which these activities can be done.

Increasing workforce capacity for prescribed fire and other fuels management treatments requires a multi-pronged approach. At Cal Poly Humboldt, the BS in Applied Fire Science and Management was created to assist in this effort. Development of this program resulted from a partnership between the Forestry, Fire, and Rangeland Management, and the Native American Studies departments with extensive consultation from local and regional practitioners across federal, state, private, Tribal, and non-profit sectors. These productive conversations were essential in creating one of the most comprehensive and novel fire education programs in the country. This degree program will provide students with an interdisciplinary education focused on developing practical knowledge and skills to become fire science or fire management professionals. The primary aim of the program is to create a better-educated workforce that recognizes the importance of fire as an ecological and cultural process. This program will teach students how to manage wildfires more effectively, to plan and implement fire and other fuels management treatments, and to mitigate the undesired impacts of wildfires.

A key component of the Applied Fire Science and Management program is the inclusion of Native American courses centered on Indigenous history, policy, and cultural practices. Cal Poly Humboldt resides on the unceded territory of the Wiyot Tribe; the broader region represents the highest Indigenous population density within California including the Hupa, Karuk, Tolowa, Wiyot, Yurok, and other Tribes. Each of these Tribes has extensive knowledge and history of cultural burning for a wide range of land stewardship goals in the region (e.g., Marks-Block et al. 2019; Halpern et al. 2022; Knight et al. 2022a). The knowledge and skills of cultural burning still exist and are the cornerstones of cultural revitalization (Kimmerer and Lake 2001; Long and Lake 2018; Marks-Block and Tripp 2021).

At Cal Poly Humboldt, we have a strong tradition of supporting Tribal students in the natural resources through programs such as the Indian Natural Resources, Science, and Engineering Program. Our students also have multiple means of engaging with Tribal communities and practices through the fire program. For instance, students participate in field trips where they meet cultural fire leaders, tour recent prescribed fire sites, and learn about Tribal burning objectives and practices. Students are also provided opportunities to participate in prescribed fire training exchanges that are co-led by local Tribes. However, much work remains in such areas as curricular development, recruiting more Indigenous students and students from other minoritized groups, and enhancing connections with Tribal communities.

Efforts to infuse the program with greater experiential learning and training opportunities for students include expanding existing partnerships with local groups and developing new partnerships. As of 2023, we have on-going or developing partnerships with Six Rivers National Forest, Mid-Klamath Restoration Council, Karuk Department of Natural Resources, Watershed Training and Research Center, Cultural Fire Management Council, California State Parks, Redwood National Park, CalFire, and the Bureau of Land Management,

Arcata Office. Given the inherent multidimensionality of the fire science and management field, the program allows greater flexibility for students to supplement the strong focus on fire science and management with individualized course selection in areas such as Botany, Geospatial Sciences, Native American Studies, Archeology, Rangeland Resources, and Wildlife Management.

Another key component of the Applied Fire Science and Management program will be the expansion of experiential learning and training opportunities in prescribed fire. In 2008, the Prescribed Fire Training Exchange (TREX)[2], a partnership between The Nature Conservancy's Fire Learning Network and the Departments of Agriculture and Interior, was created to address capacity issues related to prescribed fire implementation (Spencer, Schultz, and Hoffmann 2015). Since 2013, Cal Poly Humboldt faculty and students have regularly participated in these events to gain crucial experience and training with a diverse group of land managers, agencies, Tribes, and landowners. Beginning in 2017, a local group of private landowners and community members formed the Humboldt County Prescribed Burn Association[3]. Modeled after similar programs in the Great Plains region, this was the first of its kind in the Western U.S. In the short time since forming, it has inspired numerous other county prescribed burn associations throughout California, Oregon and Washington. This and other prescribed burn associations focus on sharing resources and people to safely implement prescribed fire projects on private lands. Humboldt County Prescribed Burn Association has also provided an excellent opportunity for Cal Poly Humboldt students to gain prescribed fire experience with a diverse set of local practitioners.

Cal Poly Humboldt can assist in increasing capacity in fire and fuels management professionals outside of traditional academic pathways by leveraging our existing educational strengths and our multifaceted connections with federal, state, Tribal, private, and non-profit partners. Through a 2022 Cal Fire grant, the Fire Resilience Institute will initiate the "Integrated Fire Education, Training, and Experience Program" by leveraging existing resources, knowledge, and capacity at Cal Poly Humboldt and through numerous well-established partnerships. This program will develop and expand our offerings related to fire and fuels education, training, and outreach in Northwestern California. Our project aims to achieve the following goals: 1) create and deploy a 24-unit fire and fuels management certificate program through Cal Poly Humboldt Extended Education Program; 2) coordinate offering 5-6 fire and fuels training courses, available to students, professionals, and community members; and 3) develop a community fire outreach and extension program. We estimate that a minimum of 100 students, professionals, and community members will participate in our integrated fire education and training program over the next four years. Through expanding the number of qualified candidates for fire and fuels management positions, the proposed program will directly facilitate many of California's Wildfire and Forest Resilience Action Plan[4] goals.

Promoting Socio-Ecological Resilience to Wildfire

Promoting resilient landscapes and communities requires adaptation to living with fire. While fire suppression will remain an important tool for mitigating impacts to communities and ecosystems, it can no longer be the primary tactic to promote longer-term landscape-level resilience (North et al. 2015). Fuel reduction treatments within the wildland-urban interface, revitalization of cultural burning practices, and strategic restoration of forests and fire to ecosystems are essential to limiting costs and the impacts of future wildfires

2. Information about the Prescribed Fire Training Exchanges and key resources are available at http://www.conservationgateway.org/ConservationPractices/FireLandscapes/HabitatProtectionandRestoration/Training/TrainingExchanges/Pages/fire-training-exchanges.aspx.

3. The Humboldt County Prescribed Burn Association website has updates and information available at https://humcopba.net/.

4. The California Wildfire and Forest Resilience Task Force website maintains updates and information, available at https://www.fire.ca.gov/media/ps4p2vck/californiawildfireandforestresilienceactionplan.pdf.

(North et al. 2021). These practices will require greater coupling of both social and ecological aspects to promote fire resilience.

Social-ecological resilience is based on the complex, non-linear dynamics of social and ecological systems (Folke 2006). Achieving this goal will have profound implications for planning and governance as we build *adaptive* capacity (Wilkinson 2012). Resilience for communities includes strategies to "cope with, adapt to, and shape change" (Imperiale and Vanclay 2016: 206). Communities are more than just resilient individuals; people work together to problem-solve, and thus community leadership and strategic action are important components to resilience (Magis 2010; Kulig and Botey 2016). Much of the existing disaster recovery literature involves case studies of communities that have experienced non-fire disasters such as hurricanes, earthquakes, and tornadoes and focuses on social and physical infrastructure (e.g., Sadri et al. 2018, Yabe et al. 2020). However, wildfire-impacted communities are uniquely able to create resilience because of the importance of human management in mitigating fire behavior and the feedbacks between social and ecological variables (Schumann et al. 2020). As identified by Berkes and Ross (2013), we envision fire-adapted community resilience as a combination of both social and ecological resilience.

As part of the Fire Resilience Institute, a newly funded project entitled "*Recovery of human communities after wildfire: building social-ecological resilience*" will begin in 2023. Principal investigators and members of the Fire Resilience Institute seek to answer the following questions: what does recovery look like in diverse communities impacted by destructive wildfires, and how are communities working towards recovery and building social-ecological resilience? We have received a Joint Fire Science Program (USDA and USDOI) grant to conduct this research in several selected case study locations in Northern California and southern Oregon. Alongside Cal Poly Humboldt graduate students, we are working with community partners (Firebrand Collective, Southern Oregon University, Slater Fire Long-term Recovery Group, Mid-Klamath Watershed Council, and Watershed Training and Research Center) engaged with wildfire recovery and resilience efforts in these locations.

As part of the new project, we plan to incorporate a participatory action research (PAR) approach. PAR is an umbrella term for a process that "begins with a research topic of importance to the community with the aim of combining knowledge and action for social change to improve community health and eliminate disparities" (Minkler et al. 2003). This approach offers a range of methods for bridging community and academic methods to knowledge generation by anchoring research more centrally within communities, sharpening the focus of project objectives, increasing the relevancy of research to partner sites, and maximizing the overall impact of the research (Chevalier and Buckles 2019). PAR complements traditional research through its unique approach in involving the community under study in guiding the research process (Wadsworth 1998), and dissemination of research findings to community stakeholders. The use of PAR has helped communities recover and adapt from floods (Meyer et al. 2018) and other natural disasters, as well as develop climate change resiliency plans (Douglas et al. 2018). In alignment with federal principles of "centering equity and environmental justice" (e.g. U.S Department of Interior 2021), incorporating community-based components is our fire resilience research in our region is especially important, as many communities most impacted by fire tend to have higher rates of poverty and are often underserved by multiple levels of government (Adams and Charnley 2020; Masri, Jin, and Wu 2022). Previous fire-related research in Northern California has also successfully engaged communities using a PAR approach to address questions about traditional ecological knowledge (Lake 2007), barriers to prescribed burning (Quinn-Davidson and Varner 2012), and food sovereignty (Sowerwine et al. 2019).

The use of a participatory action research framework to address socio-ecological resilience is important and can be effective, but this approach has its own set of challenges (Long et al. 2016). For instance, PAR requires sustained community engagement in developing and implementing research. One implication is that PAR cannot be implemented solely through student projects because of the short-term tenure of student engagement. In working to implement PAR in our fire projects, we are building on a long history of community engagement between the university and communities in our region to build institution-level connections. In conjunction with the transition to Cal Poly Humboldt, the administration recently hired a special assistant to the president who will specifically focuses on Tribal relations and community engagement. We are hopeful that this position will help bolster and improve interconnections with Tribal communities in the region. With fire, this is notably taking place with the Fire Resilience Institute, which is hiring a community fire outreach coordinator to help improve community connections. Through the institute we are also engaging and consulting with the managers and practitioners of surrounding communities to align our activities with community values and needs.

The Fire Resilience Institute-led project has the advantage of looking more cohesively at both social and ecological perspectives for communities at an intermediate stage of recovery from wildfire. Previous research on post-fire recovery has emphasized the biophysical impacts of wildfires and other climate change-related disturbances on ecological resilience (Allen 2007, Millar and Stephenson 2015). Wildfire social science has often focused on fire preparedness and mitigation (Jakes and Sturtevant 2013, McCaffrey 2015) and human community responses during and soon after wildfires (Paveglio et al. 2015, Paveglio and Edgeley 2017). There has been less attention on the longer-term recovery of communities from wildfires, especially with an emphasis on what recovery looks like for communities and how recovery after a wildfire can lead to long-term social-ecological resilience within the context of climate change. As Shumann et al. (2020) argue, wildfire recovery represents an opportunity to build transformative resilience on the landscape, moving from tactical fire recovery to responsive fire-adapted communities (see also McWethy et al. 2019).

While resilience and adaptive capacity are long-term goals, community recovery after any disaster occurs in phases from an emergency period (of rescue and relocation) to rehabilitation (when essential infrastructure is re-built) to reconstruction (when long-term physical and social needs of residents are addressed and the community attempts to rebuild in a way that increases its resilience; Colten et al. 2008). This last phase may also be termed long-term recovery, and occurs 2-3 years after a disaster, as community capacity is re-built and long-term plans are put in place (Blackman et al. 2017). Post-wildfire recovery is multifaceted, and includes rebuilding physical infrastructure; resuming municipal, social, and economic activities; and promoting psychological healing and recovery as residents struggle with "losses and prolonged distress" (Lalani et al. 2021: 2). Our research project is gathering data from three diverse sites in Northern California and Southern Oregon three to four years after they were devastated by wildfires. This work will allow us to assess both immediate post-disaster community responses and more long-term community recovery efforts in three communities that vary in size, cultural composition, and interagency organizational capacity. Rather than focusing narrowly on the "aftershock" social and economic factors (Imperiale and Vanclay 2016), we will consider the direction of recovery with several years' worth of lessons learned.

The timeline of post-disaster recovery is important: soon after disaster there may be a sense of community cohesion and working together toward a common cause (Paveglio and Edgeley 2017), a period described as a "honeymoon" phase (McGee et al. 2020). This perspective suggests that recovery efforts may lag or run into community conflicts after some time. Conversely, new actors and leaders within communities may emerge to facilitate recovery efforts and contribute to renewed social cohesion and community agency (Blackman et al. 2017). Lidskog (2018) found that nine months after a fire in Sweden, there was still increased social cohesion and people described their community as "stronger" after the fire. In one of the few studies to assess community recovery several years after the fire, Carroll et al. (2011) found that five years after the Rodeo-Chediski Fire, nearby communities had both maintained a "coming together" spirit that spurred community action, *and* had developed post-fire conflict, distrust, and disagreements. Our project will add to the existing knowledge regarding post-wildfire recovery and assess the prospects of fostering longer-term social-ecological resilience to wildfire within communities, as well as implications for other communities in the region that will experience future wildfires.

Of the limited research related to post-wildfire recovery available, Tribal perspectives are often not represented (Carroll et al. 2011) in part because of the many challenges in conducting research with Indigenous partners. To enable the inclusion of Indigenous perspectives, we will limit these barriers through incorporating lessons learned from previous projects (Long et al. 2016, Sowerwine et al. 2019), utilize guidelines when available or work with partners to assist in their development, train students, faculty, and staff in best practices, and engage with communities early to make sure expectations are clear and appropriate. These efforts will aid in better serving the project and community, while also working to establish trust and greater engagement.

Interconnecting Knowledge Systems to Foster Coexistence with Fire

Knowledge systems are the ways in which we understand and represent the world. Rather than primarily focusing on the entities involved in the process of generating and using knowledge, there is increased recognition for the need to interconnect entities across the components of knowledge systems and thus to better foster knowledge equity (Glazer 1998) or "knowledge democracy" (Veld 2010). Here we highlight one general perspective of a knowledge system with an emphasis on four components: 1) knowledge identification, 2) knowledge acquisition, 3) knowledge translation, and 4) knowledge application. Knowledge identification refers to the process of identifying gaps in knowledge or areas of uncertainty in existing knowledge. Knowledge acquisition refers to the generation of understanding through observation, experience, practice, inquiry, and other methods. There are many forms of knowledge acquisition and ways to gain important information that clarifies understanding the world and informs decision making. Knowledge translation is the dissemination and sharing of knowledge with an emphasis on providing needed information to those engaged with knowledge application, or the use of knowledge to aid decision making for management and policymaking.

An interconnected knowledge system accepts that there are multiple existing systems of knowledge and recognizes the importance of engaging across knowledge systems to find solutions while also maintaining knowledge integrity and appropriate attribution. Below, we highlight a project where we are actively using an interconnected knowledge system approach to collaboratively design and direct a monitoring and research study that aims to reintroduce fire and restore a culturally important species and landscape.

Indigenous peoples, including the Karuk and Yurok, have managed oak woodlands in lower montane regions of the Klamath Mountains since time immemorial with frequent low-intensity fire (Anderson 2005; Lake, Tripp, and Reed 2010; van Wagtendonk et al. 2018). Evidence of frequent burning includes traditional knowledge, the presence of surviving legacy trees, and fire scar and charcoal records (Metlen et al. 2018, Knight et al. 2022b). Xánthiip, or California black oak (*Quercus kelloggii*), along with other hardwoods, are important cultural, ecological, and economic species in the region (Long et al. 2016; 2017). In the long absence of fire, competition from shade-tolerant conifers such as itháriip or Douglas-fir (*Pseudotsuga menziesii*) has dramatically reduced legacy hardwood vigor. After about 100 to 150 years of fire exclusion, conifers begin to overtop black oak, leading to substantial mortality of these shade-intolerant legacy trees (Hunter and Barbour 2001; Cocking et al. 2012; Cocking et al. 2014). Reintroduction of low intensity surface fires, whether through wildfire or prescribed burning, has limited effectiveness in killing mature Douglas-fir, and may have the unfortunate outcome of further weakening or killing of legacy California black oak (Cocking et al. 2012; personal communication, Bill Tripp, Frank Lake), if mitigation measures are not taken to protect vulnerable old trees. Effective restoration of these stands often requires the mechanical removal of Douglas-fir to improve oak vigor (Devine and Harrington 2013; Kane et al. 2019). However, the long absence of fire and dense shade cast by Douglas-fir has promoted growth forms of California black oak with a low diameter-to-height ratio and excessively leaning boles that may be predisposed to windthrow, especially after thinning.

These circumstances require unique solutions to limit negative impacts while retaining California black oak and its important eco-culturally associated species. As a part of this project and others, the Fire Resilience Institute and partners will focus on a co-production model of research and employ an updated version of the Practicing Pikyav[5] guidelines for collaborative projects with the Karuk Tribe. A process that was developed and refined in part through the Karuk and University of California, Berkeley collaborative focused on food sovereignty through a participatory action research approach (Sowerwine et al. 2019).

In collaboration with the Karuk Department of Natural Resources, Mid-Klamath Watershed Council, Salmon River Restoration Council, Oregon State University, USDA Forest Service, and Southern Oregon University, members of the Humboldt Fire Resilience Institute have begun participating in the "*Black oak (xánthiip) monitoring and research project.*" This project will examine the effectiveness and impacts of specialized treatments to improve California black oak vigor and survival and to reintroduce fire and promote restoration as a part of the broader initiative of the Western Klamath Restoration Partnership. The study will take place on Ikxariyatuuyship (Offield Mountain) within the Karuk Aboriginal Territory, at a site once regularly burned in September as part of the World Renewal Ceremony (Lake, Tripp, and Reed 2010; Norgaard 2022). While the specific details of the research questions and approach are still in development, the process has been centered on Tribal cultural values and has been deliberately inclusive of multiple knowledge systems. The aim of this approach is to identify the challenges and to assure the knowledge generated will be responsive to the goals of improving social and ecological resilience in California black oak and its associated species.

Through this project each of the partners are engaged with many of the four knowledge system components. For instance, knowledge identification was initially conducted by local practitioners that identified the scope of the impacts of past management on California black oak and possible methods of improving conditions. These ideas were put forward to the larger group of partners in a series of meetings to flesh out monitoring and research approaches that were responsive to cultural and ecological perspectives. Initial monitoring protocols are being developed with continued input from many partners, with a focus on Indigenous perspectives. Knowledge acquisition will occur through both monitoring and research efforts that will be conducted by multiple partners. Inclusion of all partners early and iteratively in the process will facilitate knowledge translation and application for later stages of this project. Furthermore, this cooperative approach will also facilitate feedback into identifying subsequent knowledge needs or areas of uncertainty.

Closing Remarks

Coexistence with fire that promotes socio-ecological resilience is possible. The scope and magnitude of the wildfire-related challenges in California and other regions are complex, but substantive advances that promote inclusion and interconnection across disciplines and knowledge systems are needed. Through Cal Poly Humboldt and the Fire Resilience Institute, we are intentionally incorporating this approach. Inevitably this process will be iterative with a need to adapt over time, learn from successes, and find ways to improve on shortfalls. Our approach has been informed by many regional partners that are already actively fostering the interconnection of knowledge systems and working across boundaries to address many of the current challenges related to wildfire. We look forward to continuing to build these relationships and develop new relationships in the coming years.

Acknowledgements

The authors thank the many other partners engaged with the fire-related education, training, and research efforts at Cal Poly Humboldt and the Fire Resilience Institute. We would like to thank C. Risling Baldy and K. Reed of the Native American Studies department at Cal Poly

6. The Practicing Pikyak document and other resources are available at https://sipnuuk.karuk.us/system/files/atoms/file/ATALM17_PracticingPikyav.pdf.

Humboldt for their help and guidance in developing the new Applied Fire Science and Management program, N. Goulette (Watershed Research and Training Center), L. Latimer (Mid-Klamath Watershed Council), A. Yaeger (Slater Fire Recovery Group), T. Teutsch (Firebrand Collective), J. Trammelle (Southern Oregon University), and others for their early engagement in the socio-ecological resilience to wildfire project. We also thank the Karuk Department of Natural Resources (C. Rossier, S. Bouque, A. Tripp, C. Hillman, and others), Mid-Klamath Restoration Council (L. Latimer), Salmon River Restoration Council (K. Greenburg, A. Varner), Oregon State University (M. Krawchuck, J. Thoreson), USDA Forest Service (F. Lake), and Southern Oregon University (A. Halpern), and others for their partnership and invitation to be involved with the black oak project. Two anonymous reviewers provided helpful suggestions that improved this manuscript.

Literature Cited

Abatzoglou, John T., and A. Park Williams. 2016. "Impact of Anthropogenic Climate Change on Wildfire across Western US Forests." *Proceedings of the National Academy of Sciences* 113(42):11770–75. doi: 10.1073/pnas.1607171113.

Abrams, Jesse B., Melanie Knapp, Travis B. Paveglio, Autumn Ellison, Cassandra Moseley, Max Nielsen-Pincus, and Matthew S. Carroll. 2015. "Re-Envisioning Community-Wildfire Relations in the U.S. West as Adaptive Governance." *Ecology and Society* 20(3):art34. doi: 10.5751/ES-07848-200334.

Abrams, Jesse, Emily Jane Davis, Cassandra Moseley, and Branda Nowell. 2017. "Building Practical Authority for Community Forestry in and through Networks: The Role of Community-Based Organisations in the U.S. West: Practical Authority for Community Forestry in and through Networks." *Environmental Policy and Governance* 27(4):285–97. doi: 10.1002/eet.1765.

Adams, Mark D. O., and Susan Charnley. 2020. "The Environmental Justice Implications of Managing Hazardous Fuels on Federal Forest Lands." *Annals of the American Association of Geographers* 110(6): 1907-1935.

Allen, Craig D. 2007. "Interactions across Spatial Scales among Forest Dieback, Fire, and Erosion in Northern New Mexico Landscapes." *Ecosystems* 10(5):797–808. doi: 10.1007/s10021-007-9057-4.

Anderson, M. Kat. 2005. *Tending the Wild: Native American Knowledge and the Management of California's Natural Resources.* Berkeley, California, USA: University of California Press.

Berkes, Fikret, and Helen Ross. 2013. "Community Resilience: Toward an Integrated Approach." *Society & Natural Resources* 26(1):5–20. doi: 10.1080/08941920.2012.736605.

Bixler, R. Patrick, Shawn Johnson, Kirk Emerson, Tina Nabatchi, Melly Reuling, Charles Curtin, Michele Romolini, and J. Morgan Grove. 2016. "Networks and Landscapes: A Framework for Setting Goals and Evaluating Performance at the Large Landscape Scale." *Frontiers in Ecology and the Environment* 14(3):145–53. doi: 10.1002/fee.1250.

Blackman, Deborah, Hitomi Nakanishi, and Angela M. Benson. 2017. "Disaster Resilience as a Complex Problem: Why Linearity Is Not Applicable for Long-Term Recovery." *Technological Forecasting and Social Change* 121:89–98. doi: 10.1016/j.techfore.2016.09.018.

Butler, William H., and Bruce E. Goldstein. 2010. "The US Fire Learning Network: Springing a Rigidity Trap through Multiscalar Collaborative Networks." *Ecology and Society* 15(3):art21. doi: 10.5751/ES-03437-150321.

Butler, William H., and Courtney A. Schultz. 2019. *A New Era for Collaborative Forest Management: Policy and Practice Insights from the Collaborative Forest Landscape Restoration Program.* New York: Routledge, Abingdon, Oxon.

Carroll, Matthew S., Travis Paveglio, Pamela J. Jakes, and Lorie L. Higgins. 2011. "Nontribal Community Recovery from Wildfire Five Years Later: The Case of the Rodeo–Chediski Fire." *Society & Natural Resources* 24(7):672–87. doi: 10.1080/08941921003681055.

Cheng, Antony S., and Victoria E. Sturtevant. 2012. "A Framework for Assessing Collaborative Capacity in Community-Based Public Forest Management." *Environmental Management* 49(3):675–89. doi: 10.1007/s00267-011-9801-6.

Chevalier, Jacques. M., and Daniel J. Buckles. 2019. *Participatory Action Research: Theory and Methods for Engaged Inquiry*. New York, New York. Routledge.

Cocking, Matthew I., J. Morgan Varner, and Eric E. Knapp. 2014. "Long-Term Effects of Fire Severity on Oak-Conifer Dynamics in the Southern Cascades." *Ecological Applications* 24(1):94–107.

Cocking, Matthew I., J. Morgan Varner, and Rosemary L. Sherriff. 2012. "California Black Oak Responses to Fire Severity and Native Conifer Encroachment in the Klamath Mountains." *Forest Ecology and Management* 270:25–34. doi: 10.1016/j.foreco.2011.12.039.

Colavito Melanie M. 2017. "Utilising Scientific Information to Support Resilient Forest and Fire Management." *International Journal of Wildland Fire* 26:375–383.

Colten, Craig E., Robert W. Kates, and Shirley B. Laska. 2008. "Three Years after Katrina: Lessons for Community Resilience." *Environment: Society and Policy for Sustainable Development* 50(5):36–47.

Cumming, Graeme S. 2018. "A Review of Social Dilemmas and Social-Ecological Traps in Conservation and Natural Resource Management." *Conservation Letters* 11(1):e12376. doi: 10.1111/conl.12376.

Davis, Emily Jane, Eric M. White, Lee K. Cerveny, David Seesholtz, Meagan L. Nuss, and Donald R. Ulrich. 2017. "Comparison of USDA Forest Service and Stakeholder Motivations and Experiences in Collaborative Federal Forest Governance in the Western United States." *Environmental Management* 60(5):908–21. doi: 10.1007/s00267-017-0913-5.

Douglas, Ellen M., Kenneth M. Reardon, Matthais C. Täger. 2018. "Participatory Action Research as a Means of Achieving Ecological Wisdom within Climate Change Resiliency Planning." *Journal of Urban Management* 7(3):152-160.

Eriksen, Christine, and Don L. Hankins. 2014. "The Retention, Revival, and Subjugation of Indigenous Fire Knowledge through Agency Fire Fighting in Eastern Australia and California." *Society & Natural Resources* 27(12):1288–1303. doi: 10.1080/08941920.2014.918226.

Folke, Carl. 2006. "Resilience: The Emergence of a Perspective for Social–Ecological Systems Analyses." *Global Environmental Change* 16(3):253–67. doi: 10.1016/j.gloenvcha.2006.04.002.

Glazer, Rashi 1998. "Measuring the Knower: Towards a Theory of Knowledge Equity." *California Management Review* 40(3): 175–194.

Glenn, Evora, Laurie Yung, Carina Wyborn, and Daniel R. Williams. 2022. "Organisational Influence on the Co-Production of Fire Science Overcoming Challenges and Realising Opportunities." *International Journal of Wildland Fire* 31(4):435–48.

Gunderson, Lance H., and Crawford S. Holling. 2002. Panarchy: *Understanding Transformations in Human and Natural Systems*. Washington, DC: Island Press.

Halpern, Arielle A., Wayne P. Sousa, Frank K. Lake, Thomas J. Carlson, and Will Paddock. 2022. "Prescribed Fire Reduces Insect Infestation in Karuk and Yurok Acorn Resource Systems." *Forest Ecology and Management* 505:119768. doi: 10.1016/j.foreco.2021.119768.

Hoover, Katie. 2020. *Federal Wildfire Management: Ten-Year Funding Trends and Issues (FY2011-FY2020). Congressional Research Service Report.* R46583. Washington, DC.

Hunter, Molly E. 2016. "Outcomes of Fire Research: Is Science Used?" *International Journal of Wildland Fire* 25(5):495–504.

Hunter, John C., and Michael G. Barbour. 2001. "Through-Growth by Pseudotsuga menziesii: A Mechanism for Change in Forest Composition without Canopy Gaps." *Journal of Vegetation Science* 12(4):445–52. doi: 10.2307/3236996.

Hunter, Molly E., Melanie M. Colavito, and Vita Wright. 2020. "The Use of Science in Wildland Fire Management: A Review of Barriers and Facilitators." *Current Forestry Reports* 6(4):354–67. doi: 10.1007/s40725-020-00127-2.

Imperiale, Angelo J., and Frank Vanclay. 2016. "Experiencing Local Community Resilience in Action: Learning from Post-Disaster Communities." *Journal of Rural Studies* 47:204–19. doi: 10.1016/j.jrurstud.2016.08.002.

Jakes, Pamela J., and Victoria Sturtevant. 2013. "Trial by Fire: Community Wildfire Protection Plans Put to the Test." *International Journal of Wildland Fire* 22(8):1134. doi: 10.1071/WF12156.

Kane, Jeffrey M., Eamon A. Engber, and John E. McClelland. 2019. "Effectiveness and Impacts of Girdling Treatments in a Conifer-Encroached Oregon White Oak Woodland." *Forest Ecology and Management* 447:77–86. doi: 10.1016/j.foreco.2019.05.059.

Kelly, Erin C., Susan Charnley, and J. T. Pixley. 2019. "Polycentric Systems for Wildfire Governance in the Western United States." *Land Use Policy* 14.

Kimmerer, Robin W., and Frank K. Lake. 2001. "The Role of Indigenous Burning in Land Management." *Journal of Forestry* 99(11):36–41.

Knight, Clarke A., Ryan E. Tompkins, Jonathan A. Wang, Robert York, Michael L. Goulden, and John J. Battles. 2022a. "Accurate Tracking of Forest Activity Key to Multi-Jurisdictional Management Goals: A Case Study in California." *Journal of Environmental Management* 302:114083. doi: 10.1016/j.jenvman.2021.114083.

Knight, Clarke A., Lysanna Anderson, M. Jane Bunting, Marie Champagne, Rosie M. Clayburn, Jeffrey N. Crawford, Anna Klimaszewski-Patterson et al. 2022b. "Land Management Explains Major Trends in Forest Structure and Composition over the Last Millennium in California's Klamath Mountains." *Proceedings of the National Academy of Sciences* 119(12):e2116264119. doi: 10.1073/pnas.2116264119.

Kobziar, Leda N., Monique E. Rocca, Christopher A. Dicus, Chad Hoffman, Neil Sugihara, Andrea E. Thode, J. Morgan Varner, and Penelope Morgan. 2009. "Challenges to Educating the Next Generation of Wildland Fire Professionals in the United States." *Journal of Forestry* 107(7):339-345.

Kocher, Susan D., Eric Toman, Sarah F. Trainor, Vita Wright, Jennifer S. Briggs, Charles P. Goebel, Eugénie M. MontBlanc, Annie Oxarart, Donna L. Pepin, Toddi A. Steelman, Andrea Thode, and Thomas A. Waldrop. 2012. "How Can We Span the Boundaries between Wildland Fire Science and Management in the United States?" *Journal of Forestry* 110(8):421–28. doi: 10.5849/jof.11-085.

Kulig, Judith, and Anna Pujadas Botey. 2016. "Facing a Wildfire: What Did We Learn about Individual and Community Resilience?" *Natural Hazards* 82(3):1919–29. doi: 10.1007/s11069-016-2277-1.

Lake, Frank K. 2007. "Traditional Ecological Knowledge to Develop and Maintain Fire Regimes in Northwestern California, Klamath-Siskiyou Bioregion: Management and Restoration of Culturally Significant Habitats." PhD dissertation, Oregon State University, Corvallis.

Lake, Frank K., William Tripp, and Ron Reed. 2010. "The Karuk Tribe, Planetary Stewardship, and World Renewal on the Middle Klamath River, California." *Bulletin of the Ecological Society of America* 91:147-149.

Lebel, Louis, Jesse B. Manuta, and Po Garden. 2011. "Institutional Traps and Vulnerability to Changes in Climate and Flood Regimes in Thailand." *Regional Environmental Change* 11(1):45–58. doi: 10.1007/s10113-010-0118-4.

Lidskog, Rolf. 2018. "Invented Communities and Social Vulnerability: The Local Post-Disaster Dynamics of Extreme Environmental Events." *Sustainability* 10(12):4457. doi: 10.3390/su10124457.

Long, Jonathan W., and Frank K. Lake. 2018. "Escaping Social-Ecological Traps through Tribal Stewardship on National Forest Lands in the Pacific Northwest, United States of America." *Ecology and Society* 23(2):art10. doi: 10.5751/ES-10041-230210.

Long, Jonathan W., M. Kat Anderson, Lenya Quinn-Davidson, Ron W. Goode, Frank K. Lake, and Carl N. Skinner. 2016. *Restoring California Black Oak Ecosystems to Promote Tribal Values and Wildlife.* PSW-GTR-252. Albany, CA: U.S. Department of

Agriculture, Forest Service, Pacific Southwest Research Station. doi: 10.2737/PSW-GTR-252.

Long, Jonathan W., Ron W. Goode, Raymond J. Gutteriez, Jessica J. Lackey, and M. Kat Anderson. 2017. "Managing California Black Oak for Tribal Ecocultural Restoration." *Journal of Forestry* 115(5):426–34. doi: 10.5849/jof.16-033.

Magis, Kristen. 2010. "Community Resilience: An Indicator of Social Sustainability." *Society & Natural Resources* 23(5):401–16. doi: 10.1080/08941920903305674.

Maletsky, Lisa D., William P. Evans, Loretta Singletary, and Lorie L. Sicafuse. 2018. "Joint Fire Science Program (JFSP) Fire Science Exchange Network: A National Evaluation of Initiative Impacts." *Journal of Forestry* 116(4):328–35. doi: 10.1093/jofore/fvy009.

Mann, Michael L., Enric Batllori, Max A. Moritz, Eric K. Waller, Peter Berck, Alan L. Flint, Lorraine E. Flint, and Emmalee Dolfi. 2016. "Incorporating Anthropogenic Influences into Fire Probability Models: Effects of Human Activity and Climate Change on Fire Activity in California" *PLOS ONE* 11(4):e0153589. doi: 10.1371/journal.pone.0153589.

Marks-Block, Tony, Frank K. Lake, Rebecca Bliege Bird, and Lisa M. Curran. 2021. "Revitalized Karuk and Yurok Cultural Burning to Enhance California Hazelnut for Basketweaving in Northwestern California, USA." *Fire Ecology* 17(1):6. doi: 10.1186/s42408-021-00092-6.

Marks-Block, Tony, and William Tripp. 2021. "Facilitating Prescribed Fire in Northern California through Indigenous Governance and Interagency Partnerships." *Fire* 4(3):37. doi: 10.3390/fire4030037.

Masri, Shahir, Yufang Jin, and Jun Wu. 2022. "Compound Risk of Air Pollution and Heat Days and the Influence of Wildfire by SES across California, 2018–2020: Implications for Environmental Justice in the Context of Climate Change." *Climate* 10(10):145. doi: 10.3390/cli10100145.

McCaffrey, Sarah. 2015. "Community Wildfire Preparedness: A Global State-of-the-Knowledge Summary of Social Science Research." *Current Forestry Reports* 1(2):81–90. doi: 10.1007/s40725-015-0015-7.

McWethy, David B., Tania Schoennagel, Philip E. Higuera, Meg Krawchuk, Brian J. Harvey, Elizabeth C. Metcalf, Courtney Schultz, et al. 2019. "Rethinking Resilience to Wildfire." *Nature Sustainability* 2(9):797–804. doi: 10.1038/s41893-019-0353-8.

Melvin, Mark A. 2018. 2018 P*rescribed Fire Use Survey Report. Technical Report.* 03–18. Coalition of Prescribed Fire Councils, Inc.

Metlen, Kerry L., Carl N. Skinner, Derek R. Olson, Clint Nichols, and Darren Borgias. 2018. "Regional and Local Controls on Historical Fire Regimes of Dry Forests and Woodlands in the Rogue River Basin, Oregon, USA." *Forest Ecology and Management* 430:43–58. doi: 10.1016/j.foreco.2018.07.010.

Meyer, Michelle A., Marccus M. Hendricks, Galen D. Newman, Jamie H. Masterson, John T. Cooper, Garett Sansom, Nasir Gharaibeh, et al. 2018. "Participatory Action Research: Tools for Disaster Resilience Education." *International Journal of Disaster Resilience in the Built Environment* 9(4-5):402–419.

Millar, Constance I., and Nathan L. Stephenson. 2015. "Temperate Forest Health in an Era of Emerging Megadisturbance." *Science* 349(6250):823–26. doi: 10.1126/science.aaa9933.

Minkler, M., Blackwell, A.G., Thompson, M., Tamir, H. 2003. "Community-based Participatory Research: Implications for Public Health Funding." *American Journal of Public Health* 93(8):1210–1213.

Moritz, Max A., Enric Batllori, Ross A. Bradstock, A. Malcolm Gill, John Handmer, Paul F. Hessburg, Justin Leonard, et al. 2014. "Learning to Coexist with Wildfire." *Nature* 515(7525):58–66. doi: 10.1038/nature13946.

Nair, Sreeja, and Michael Howlett. 2016. "From Robustness to Resilience: Avoiding Policy Traps in the Long Term." *Sustainability Science* 11(6):909–17. doi: 10.1007/s11625-016-0387-z.

Norgaard, Kari Marie. 2022. "The Politics of Fire and the Social Impacts of Fire Exclusion on the Klamath." *Humboldt Journal of Social Relations* (36):26.

Norström, Albert V., Christopher Cvitanovic, Marie F. Löf, Simon West, Carina Wyborn, Patricia Balvanera, Angela T. Bednarek, et al. 2020. "Principles for Knowledge Co-Production in Sustainability Research." *Nature Sustainability* 3(3):182–90. doi: 10.1038/s41893-019-0448-2.

North, Malcolm, Brandon M. Collins, and Scott Stephens. 2012. "Using Fire to Increase the Scale, Benefits, and Future Maintenance of Fuels Treatments." *Journal of Forestry* 110(7):392–401.

North, Malcolm. P., Scott L. Stephens, Brandon M. Collins, James K. Agee, Gregory H. Aplet, Jerry F. Franklin, and Peter Z. Fulé. 2015. "Reform Forest Fire Management." *Science* 349(6254):1280–81.

North, Malcolm. P., Robert A. York, Brandon M. Collins, Matthew D. Hurteau, Gavin M. Jones, Eric E. Knapp, Leda Kobziar, et al. 2021. "Pyrosilviculture Needed for Landscape Resilience of Dry Western United States Forests." *Journal of Forestry* 119(5):520-544. doi: 10.1093/jofore/fvab026.

Noss, Reed F., Jerry F. Franklin, William L. Baker, Tania L. Schoennagel, and Peter B. Moyle. 2006. "Managing Fire-prone Forests in the Western United States." *Frontiers in Ecology and the Environment* 4(9):481–87.

Parks, Sean A., and John T. Abatzoglou. 2020. "Warmer and Drier Fire Seasons Contribute to Increases in Area Burned at High Severity in Western US Forests From 1985 to 2017." *Geophysical Research Letters* 47(22):e2020GL089858. doi: 10.1029/2020GL089858.

Paveglio, Travis B., Cassandra Moseley, Matthew S. Carroll, Daniel R. Williams, Emily Jane Davis, and A. Paige Fischer. 2015. "Categorizing the Social Context of the Wildland Urban Interface: Adaptive Capacity for Wildfire and Community 'Archetypes.'" *Forest Science* 61(2):298–310. doi: 10.5849/forsci.14-036.

Paveglio, Travis, and Catrin Edgeley. 2017. "Community Diversity and Hazard Events: Understanding the Evolution of Local Approaches to Wildfire." *Natural Hazards* 87(2):1083–1108. doi: 10.1007/s11069-017-2810-x.

Quinn-Davidson, Lenya N., and J. Morgan Varner. 2012. "Impediments to Prescribed Fire across Agency, Landscape and Manager: An Example from Northern California." *International Journal of Wildland Fire* 21(3):210–218. doi: 10.1071/WF11017.

Quigley, Thomas M., Richard W. Haynes, and Russell T. Graham. 1996. *Integrated Scientific Assessment for Ecosystem Management in the Interior Columbia Basin, and Portions of the Klamath and Great Basins. General Technical Report.* PNW-GTR-382. Portland, Oregon: USDA Forest Service Pacific Northwest Research Station.

Ryan, Kevin C., Eric E. Knapp, and J. Morgan Varner. 2013. "Prescribed Fire in North American Forests and Woodlands: History, Current Practice, and Challenges." *Frontiers in Ecology and the Environment* 11(s1):e15–24. doi: 10.1890/120329.

Sadri, Arif Mohaimin, Satish V. Ukkusuri, Seungyoon Lee, Rosalee Clawson, Daniel Aldrich, Megan Sapp Nelson, Justin Seipel, and Daniel Kelly. 2018. "The Role of Social Capital, Personal Networks, and Emergency Responders in Post-Disaster Recovery and Resilience: A Study of Rural Communities in Indiana." *Natural Hazards* 90(3):1377–1406. doi: 10.1007/s11069-017-3103-0.

Schultz, Courtney A., Sarah M. McCaffrey, and Heidi R. Huber-Stearns. 2019. "Policy Barriers and Opportunities for Prescribed Fire Application in the Western United States." *International Journal of Wildland Fire* 28(11):874–884. doi: 10.1071/WF19040.

Schumann, Ronald L., Miranda Mockrin, Alexandra D. Syphard, Joshua Whittaker, Owen Price, Cassandra Johnson Gaither, Christopher T. Emrich, and Van Butsic. 2020. "Wildfire Recovery as a 'Hot Moment' for Creating Fire-Adapted Communities." *International Journal of Disaster Risk Reduction* 42:101354. doi: 10.1016/j.ijdrr.2019.101354.

Shuman, Jacquelyn K., Jennifer K. Balch, Rebecca T. Barnes, Philip E. Higuera, Christopher I. Roos, Dylan W. Schwilk, E. Natasha Stavros et al. 2022. "Reimagine Fire Science for the Anthropocene." *PNAS Nexus* pgac115. doi: 10.1093/pnasnexus/pgac115.

Sowerwine, Jennifer, Daniel Sarna-Wojcicki, Megan Mucioki, Lisa Hillman, Frank Lake, and Edith Friedman. 2019. "Enhancing Food Sovereignty: A Five-Year Collaborative Tribal-University Research and Extension Project in California and Oregon." *Journal of Agriculture, Food Systems, and Community Development* 9(B):167–190. doi: 10.5304/jafscd.2019.09B.013.

Spencer, Andrew G., Courtney A. Schultz, and Chad M. Hoffman. 2015. "Enhancing Adaptive Capacity for Restoring Fire-Dependent Ecosystems: The Fire Learning Network's Prescribed Fire Training Exchanges." *Ecology and Society* 20(3):art38. doi: 10.5751/ES-07847-200338.

Steel, Zachary L., Hugh D. Safford, and Joshua H. Viers. 2015. "The Fire Frequency-Severity Relationship and the Legacy of Fire Suppression in California Forests." *Ecosphere* 6(1):art8. doi: 10.1890/ES14-00224.1.

Steelman, Toddi. 2016. "U.S. Wildfire Governance as Social-Ecological Problem." *Ecology and Society* 21(4):art3. doi: 10.5751/ES-08681-210403.

Stephens, Scott L., and Lawrence W. Ruth. 2005. "Federal Forest-Fire Policy in the United States." *Ecological Applications* 15(2):532–42. doi: 10.1890/04-0545.

Syphard, Alexandra D., Jon E. Keeley, and John T. Abatzoglou. 2017. "Trends and Drivers of Fire Activity Vary across California Aridland Ecosystems." *Journal of Arid Environments* 144:110–22. doi: 10.1016/j.jaridenv.2017.03.017.

US Department of Agriculture, Forest Service and State of California. 2020. Agreement for Shared Stewardship of California's Forests and Rangelands.

US Department of Agriculture, Forest Service. 2012. National Forest System Land Management Planning Final Rule and Record of Decision. 36 CFR Part 219. *Federal Register* 77(68): 21162-21276. http://www.fs.usda.gov/Internet/FSE_DOCUMENTS/stelprdb5362536.pdf.

van Wagtendonk, J. W., Neil G. Sugihara, Scott L. Stephens, Andrea E. Thode, Kevin E. Shaffer, and Jo Ann Fites-Kaufman, eds. 2018. *Fire in California's Ecosystems*. Second. Berkeley: University of California Press.

Wadsworth, Y. J. 1998. *What is Participatory Action Research? Action Research International*. Melbourne, Australia: Southern Cross Institute for Action Research.

Westerling, Anthony L. 2016. "Increasing Western US Forest Wildfire Activity: Sensitivity to Changes in the Timing of Spring." *Philosophical Transactions of the Royal Society B: Biological Sciences* 371(1696):20150178. doi: 10.1098/rstb.2015.0178.

Wilkinson, Cathy. 2012. "Social-Ecological Resilience: Insights and Issues for Planning Theory." *Planning Theory* 11(2):148–69. doi: 10.1177/1473095211426274.

Wurtzebach, Zachary, and Courtney Schultz. 2016. "Measuring Ecological Integrity: History, Practical Applications, and Research Opportunities." *BioScience* 66(6):446–57. doi: 10.1093/biosci/biw037.

Yabe, Takahiro, Kota Tsubouchi, Naoya Fujiwara, Yoshihide Sekimoto, and Satish V. Ukkusuri. 2020. "Understanding Post-Disaster Population Recovery Patterns." *Journal of The Royal Society Interface* 17(163):20190532. doi: 10.1098/rsif.2019.0532.

Yaffee, Steven L. 1996. "Ecosystem Management in Practice: The Importance of Human Institutions." *Ecological Applications* 6(3):724–27. doi: 10.2307/2269472

HJSR
50TH ANNIVERSARY EDITION

Transformative Sea-Level Rise Research and Planning: Establishing a University, Tribal, and Community Partnership for a Resilient California North Coast

Laurie Richmond *<laurie.richmond@humboldt.edu> (Cal Poly Humboldt; California Sea Grant),* ***Jeff Anderson*** *(Northern Hydrology),* ***Josephine Archibald*** *(Cal Poly Humboldt),* ***Alec Brown*** *(Cal Poly Humboldt),* ***Adam Canter*** *(Wiyot Tribe),* ***Eileen Cashman*** *(Cal Poly Humboldt),* ***Joice Chang*** *(Cal Poly Humboldt),* ***Clancy De Smet*** *(Caltrans),* ***Yvonne Everett*** *(Cal Poly Humboldt),* ***James Graham*** *(Cal Poly Humboldt),* ***Nayre Herrera*** *(Cal Poly Humboldt),* ***Bente Jansen*** *(California Coastal Commission),* ***Jennifer Kalt*** *(Humboldt Baykeeper; Cal Poly Humboldt),* ***Aldaron Laird*** *(Cal Poly Humboldt Sea-Level Rise Institute),* ***Lonyx Landry*** *(Cal Poly Humboldt),* ***Daniel Lipe*** *(Cal Poly Humboldt),* ***Bonnie Ludka*** *(Cal Poly Humboldt),* ***José R. Marín Jarrín*** *(Cal Poly Humboldt),* ***Jennifer Marlow*** *(Cal Poly Humboldt),* ***Kristen Orth-Gordinier*** *(GHD, Inc.),* ***Jason R. Patton*** *(Cal Poly Humboldt),* ***Frank Shaughnessy*** *(Cal Poly Humboldt),* ***Alyssa Suarez*** *(Cal Poly Humboldt),* ***Alexandra Toyofuku*** *(Cal Poly Humboldt),* ***Amelia R. Vergel de Dios*** *(Cal Poly Humboldt),* ***Hilanea Wilkinson*** *(Wiyot Tribe)*

Abstract

Sea-level rise (SLR) is and will continue to be a pressing issue in the rural, North Coast region of California, especially since nearby Wigi (or Humboldt Bay) is experiencing one of the fastest rates of relative SLR on the U.S. West Coast. In this paper, we argue that SLR presents a transformative opportunity to rekindle environmental relationships and reshape the future of the California North Coast and beyond. As the preeminent higher education institution of the region, Cal Poly Humboldt has the responsibility to be a leader in education, research, and planning for climate resilience. We describe efforts of the Cal Poly Humboldt Sea Level Rise Institute to establish a university-Tribal-community partnership that braids together different approaches and ways of knowing to develop research and planning that supports a resilient California North Coast. Since Wigi is projected to experience the effects of SLR sooner than the rest of the state, the North Coast region is poised to act as an incubator for new ideas and solutions, including Indigenous knowledge systems, and to play a role in influencing equitable, resilient, and transformative SLR adaptation processes in other parts of the state and the world. This will require developing programming and expertise in specific disciplinary areas, but, more importantly, will require the development of opportunities and spaces for various disciplines, ways of knowing, and sectors (e.g. Tribal nations, academia, government, NGOs, private companies, and community groups) to converge and bring the best of what they have to address climate-induced challenges and opportunities.

Keywords: sea-level rise, climate resilience, adaptation, transformation

A shift in thinking would open up the potential for climate change research to be a transformative moment, especially if democratized and inclusive of multiple forms of knowledge.

— *Hardy, Milligan, and Heynen 2017:71*

Hopefully all of us can work together [. . .] and we can help battle climate change and try to adapt.

— *Hilanea Wilkinson, Citizen of the Wiyot Tribe, Humboldt Bay Symposium 2021*

Sea-level rise (SLR) is an outcome of the global gclimate crisis. Recent studies estimate that up to 171 million people across the globe face risk of coastal flooding, but those figures are expected to jump to 204 million by 2050 and 253 million by 2100 (Plumer 2020). Even if global heating is limited to well below 2 °C, sea level will continue to rise. Global mean sea level has risen 20-23 cm since 1880 and the rate of rise has been accelerating. Since 1850, anthropogenic climate change has been the primary factor contributing to global sea level rise (Jevrejeva, Grinsted, and Moore 2009; Church et al. 2013). This rise is mostly attributed to the thermal expansion of seawater as it warms and meltwater from glaciers and ice sheets (Frederikse et al. 2020; Lindsey 2022; Oppenheimer et al. 2022). SLR is projected to have a large effect on the state, economy, and people of California (OPC 2018). The state, the nation, and the entire globe are in search of strategies and solutions to address this issue. SLR is and will continue to be a pressing issue in the rural North Coast region of California where Cal Poly Humboldt is located. For example, nearby Wigi[1] (or Humboldt Bay), which is part of the ancestral and present day territory of the Wiyot people, is experiencing the fastest rate of relative SLR on the U.S. West Coast (Anderson 2018). SLR has the potential to inundate critical infrastructure, residential, agricultural, industrial, and ecologically and culturally significant coastal lands in the region (Laird 2018b). These changes could lead to severe social, cultural, economic, and environmental harm without effective adaptive responses. Despite these risks, SLR presents people in our region with an opportunity to work together and envision a more just and resilient future.

This paper has been developed by academic, Tribal, community, and government members of the Cal Poly Humboldt Sea Level Rise Institute (SLRI), which formed as an interdisciplinary initiative in 2018. The goals of this paper are to (1) provide readers with an overview of SLR in the North Coast region, (2) provide a description of how different ways of thinking influence our approaches to SLR and some of the work that has been done locally to date, and (3) provide a vision for how Cal Poly Humboldt, in concert with Tribal, community, government, and private sector partners, can develop programming and approaches related to equitable, resilient, and transformative SLR education, research, planning, and action.

As we describe our vision for local SLR work into the future, we draw from the following definitions:

Equitable: Many scholars and practitioners have begun exploring the concept of equity in relation to marine and coastal systems or climate adaptation (e.g. Mohnot, Bishop, and Sanchez 2019; Bennett et al. 2021; Swanson 2021; Ajibade et al. 2022; Bibby and Carter 2022; OPC 2022). Equity has been defined as a "process that leads to the intended outcome of justice, which necessitates the reckoning, remedying, and prevention of sys-

1. Throughout the paper we will use term Wigi which is the Wiyot name for Humboldt Bay.

temic injustices" (Bibby and Carter 2022:11). We define equitable SLR planning and research as that which seeks justice and considers how SLR and SLR adaptation processes will interact with differences based on race, class, gender, sexuality, and ability among others. It prioritizes action in disadvantaged or frontline communities. Equitable planning processes seek to address historical inequities and are inclusive, incorporating Tribal leadership, community involvement, and multiple ways of knowing. Ajibade et al. (2022:8) further define just SLR adaptation processes as those that incorporate "self-determination, robust planning, collaborative decision making, land rights,...and attention to intersecting justice(s)."

Resilient: People are part of the natural world and we depend on ecosystems for survival. Resilience describes the ability of interconnected social-ecological systems to adapt to change and thrive in dynamic environments (Resilience Alliance 2023). Beatley (2009) defines coastal resilience specifically as composed of four planning dimensions: ecological, social, and economic resilience and resilience of land use and the built environment—all of which must be weighed and addressed in effective coastal planning. SLR planning for resilience is multi-dimensional, considering the longevity and effectiveness of potential adaptation measures in terms of protecting communities, local economies, infrastructure, ecosystems, and public safety in the face of dynamic coastal change. Planning for resilience considers the potential impacts of SLR adaptation actions on ecosystems, habitats, and species and seeks to develop solutions that address environmental impacts from SLR, such as the loss of coastal habitats from rising waters. As an example, planning for resilience may not prioritize the building of levees or sea walls around locations when those strategies could have significant negative environmental consequences (Dugan et al. 2011) or if those areas are projected to flood from rising groundwater just a short time after the structures are completed (Rotzoll and Fletcher 2013; Pierre-Louis 2021).

Transformative: Pelling (2011:83) introduces the concept of climate change "adaptation as transformation." He argues that "it is too easy to see adaptation as a narrowly defensive task—protecting core assets or functions from the risk of climate change," (Pelling 2011:i) and instead invites a shift in thinking toward viewing climate change adaptation as "an opportunity for social reform, for the questioning of values that drive inequalities in development and our unsustainable relationship with the environment" (Pelling 2011:3). For us, transformative SLR work is that which considers opportunities to disrupt the status quo and pursues forms of development and governance that overcome historic injustices, center equity, and reconstitute our relationship with the natural world. Transformative climate resilience work also invites opportunities for climate-induced challenges to change the way institutions (like Cal Poly Humboldt) and disciplines operate, how they seek and communicate knowledge, how they collaborate, how they make decisions, and how they wield, share, and re-center power.

Braiding Disciplines and Ways of Knowing

There has been much scholarship dedicated to the challenge of developing research and solutions to complex environmental issues that spill across disciplinary, sectoral, and knowledge boundaries (e.g. Rittel and Webber 1973; Cheng 2006; Weber, Lach, and Steel 2017). Scholars in academia have for decades recognized the need to bring together multiple disciplines in order to address environmental challenges, with a focus on bridging social and environmental approaches. The notion of interdisciplinarity or its cousin transdisciplinarity has "become a widespread mantra for research" (Klein 2008:S116) within many branches of academia. Scholarship reveals that inter- or trans-disciplinary environmental work has achieved many important breakthroughs, but has also run into challenges related to overcoming disciplinary and ontological divides and incorporating (or failing to incorporate) perspectives from communities outside of academia (Polk 2014, 2015; Gaziulusoy et al. 2016). Transdisciplinary scholarship can be a source of inspiration for SLR work locally as it typically involves the incorporation of multiple disciplines and knowledges, it is driven by societal and community needs, and it "presumes theoretical and methodological transformation of each discipline will take place through the process of research" (Gaziulusoy et al. 2016:56). Still, the centering of the word "discipline" in these terms forefronts the work as predominantly drawn from academic spaces and modes of thinking.

Starting around the late 1980s, scholars from the environmental fields began to increasingly acknowledge the existence and importance of multiple knowledge-types or ways of knowing the natural world (Bateson 1979; Berkes 1999; Barnhardt and Kawagley 2005). Academics defined the term local ecological knowledge (LEK) to describe the place-based and experiential knowledge that individuals or communities develop through their connections to ecosystems. Scholars also proposed the term Traditional Ecological Knowledge (TEK) to describe the environmental knowledge generated by Indigenous groups through their connection to and stewardship of environments over many generations (Johnson 1992; Berkes 1999; Berkes, Colding, and Folke 2000; Ramos 2021). Many of these same scholars have called for the "integration" of knowledge types to understand and address complex environmental issues (Berkes 1999, Reid et al. 2006). However, these integration approaches have drawn critique. Scholars point out that the term TEK was developed by Western scientists and it does not necessarily originate from or reflect the ways indigenous people describe their worldviews (McGregor 2005; Reed 2022). The notion of TEK has also been critiqued for being reductive—seeking to isolate environmental information or observations from the broader cultural and ethical context (Kimmerer 2000; Agrawal 2002). Finally, scholars have critiqued the entire project of "knowledge integration" by environmental scientists for upholding rather than overcoming existing power dynamics where Western modes of understanding maintain primacy. They argue that in these integration practices, Western scientists continue to lead the effort and Indigenous and community-based forms of knowledge are required to be communicated and packaged in a Western science format rather than shifting Western science to better fit within the context of other ways of knowing (Nadasdy 1999; Agrawal 2002).

Scholars and practitioners have come up with many terms to describe the knowledge, wisdom, and worldviews that Indigenous groups possess related to environmental systems, including: Indigenous Knowledge, Indigenous Science, Traditional Ecological/Environmental Knowledge (TEK), and Native Science among others (Johnson [1992] 1998; Cajete 2000; Menzies 2006; Lipe 2013; Hernandez 2022). For the purposes of this paper, we use the term Indigenous Knowledges (IKs), since it foregrounds its relation to indigenous people; it is broad and does not require the separation of environmental and cultural elements; and it is plural, signaling the multiplicity of ways different Indigenous groups relate to the world. A more detailed discussion about history, definitions, and implications of these different terms can be found in Reed (2022) and another paper in this special issue (Baldy, Begay, and Reed 2023)..

New academic frameworks have attempted to move past some of the dynamics from inter/trans disciplinary and knowledge integration approaches. The concept of "civic science" highlights doing academic work in partnership with community, weaving together "science, story, and community" (The Land Institute 2023). Additionally, the notion of "convergent science" aims to address the need for a more coherent and integrated approach to hazard risk reduction. Peek et al. (2020:2) defined convergent science as:

> [a]n approach to knowledge production and action that involves diverse teams working together in novel ways—transcending disciplinary and organizational boundaries—to address vexing social, economic, environmental, and technical challenges in an effort to reduce disaster losses and promote collective well-being.

Many of these academic frameworks, while doing much to move environmental research in more holistic and just directions, tend to present their ideas as wholly new, when in fact Indigenous worldviews and knowledges are based upon the seamless incorporation of social, cultural, ethical, and environmental elements. When thinking about a vision for SLR work at Cal Poly Humboldt, we sought an approach that could acknowledge Indigenous practitioners as the true leaders in holistic environmental stewardship. We also sought an approach that confronts entrenched power dynamics and forefronts Indigenous knowledges and leadership, while also acknowledging the important contributions that academic and scientific knowledge can bring to understanding and addressing SLR.

We have chosen to draw inspiration from the allegory of the braid as described by Robin Wall Kimmerer (2013) in her seminal work *Braiding Sweetgrass*. In the book she describes how, in braiding sweetgrass, individual "strands once separated are rewoven into a new whole" (Kimmerer 2013:256). This allegory might extend locally to processes of weaving with plant materials to create baskets, regalia, and other cultural items. Weaving remains an important part of the culture of many California North Coast Tribes including the Wiyot Tribe (CIBA 2022). Indigenous weavers in this region utilize materials from many local plants including spruce root, willow, alder, hazel, beargrass, and ferns (Landry, personal communication 2022). The notion of braiding or weaving could also provide a useful framework for how Cal Poly Humboldt might move forward to fully realize its potential as an institution to address climate issues like SLR. We at the SLRI envision building a space where different strands of thinking—whether they be different ways of knowing or academic disciplines—can be woven or braided together to create a strengthened, holistic, and inclusive approach.

This paper begins with some background information about SLR in Wigi and the North Coast region, then through separate sections, describes the approaches from five of the many different strands or ways of thinking connected to SLR concerns in the region: Indigenous Knowledges and Leadership; Geosciences; Marine and Coastal Science; Engineering; and Social Science, Policy, Law, and Planning. Finally, we describe the work of braiding those strands together into a more holistic vision for equitable, resilient, and transformative SLR practice. We also present a set of suggestions for how Cal Poly Humboldt, with many partners, can grow and develop to take on the responsibility of becoming a leader in SLR, climate, and community resilience work.

Sea-Level Rise and the California North Coast Region

The California North Coast region—defined roughly as the coastal areas in Mendocino, Humboldt, and Del Norte counties—is remote, lower-income, and less developed than the more urban coastline in both Southern California and around the Bay Area. There is a strong Tribal presence in the region; over a dozen Tribal Nations and Tribal organizations have citizens whose ancestral territory includes the coastline of the North Coast. The region is home to three major ports and numerous coastal communities. As the only major academic institution in the region, Cal Poly Humboldt has the responsibility to support and collaborate with Tribes and communities throughout this region as they face SLR.

Situated in the middle of the region is Wigi. Given that Wigi is experiencing the fastest rate of SLR in California, much of the initial SLR work has focused there. The Wiyot People have occupied the lands surrounding Wigi since time immemorial. Prior to settler arrival, the Wiyot "occupied a string of villages that encircled" Wigi. Many of these village sites were in coastal wetlands and the Wiyot people used the bay for sustenance, travel, and cultural practices (Rohde 2020:22). White settlement was devastating to the Wiyot people, notably through a series of massacres in 1860 which reduced the population from 1,500-2,000 to just 100 by 1910 (Wiyot Tribe 2022). Despite this devastation, the Wiyot people have continued to occupy this region and maintain their cultural worldviews. Descendants of the Wiyot People are members of several present-day Tribal Nations including the Wiyot Tribe, Bear River Band of the Rohnerville Rancheria, Blue Lake Rancheria, and Trinidad Rancheria. The Wiyot Tribe alone is now over 600 members strong. The Wiyot Tribe states that, "Wiyot culture remains living and dynamic to this day" and the Tribe's Cultural Heritage Center recognizes "culture as a dynamic process and the Wiyot as a living people." The Heritage Center maintains a goal to "treasure the past, enrich the present, and meet the challenges of the future" (Wiyot Tribe 2022).

In many ways, this colonial history has set the stage for the SLR issues Wigi now faces. In addition to the devastation that white settlement left on Indigenous inhabitants of the region, the settlement period also contributed significant alterations to the Wigi landscape. In order to create grazing and agricultural lands, settlers built a series of dikes and levees around the shores of Wigi (Rohde 2020). These engineered shorelines cut off coastal wetlands from the bay and resulted in the loss of nearly 90% of the bay's saltwater wetlands and marshes (Barnhart, Boyd, and Pequegnat 1992). The legacy of this modification created the conditions for the SLR and flooding challenges facing all communities around Wigi today. A 2013 inventory shows that 75% of Wigi's shoreline is artificial, much of which is decaying and at risk of breaching (Laird, Powell, and Anderson 2013; Laird and Powell 2020). The parts of Wigi projected to be inundated with SLR overlap with the historic location of salt marsh habitat; meaning that SLR, without intervention, could return Wigi close to its pre-colonization size. Figure 1 shows maps of the historic salt marsh (a) and projected areas for inundation with 0.5 and 1.0 meters of SLR (b).

Since the construction of the artificial shoreline, critical infrastructure was developed and now exists in flood-prone areas. Rising seas will need to be met by substantial restoration, adaptation, and/or relocation to avoid the disruption of services and to regain ecological functioning of historic salt marsh areas. The Wigi shoreline is governed by a patchwork of Tribal, city, county, state, federal and regional government entities, all of whom will need to coordinate for effective adaptation (Humboldt County 2018).

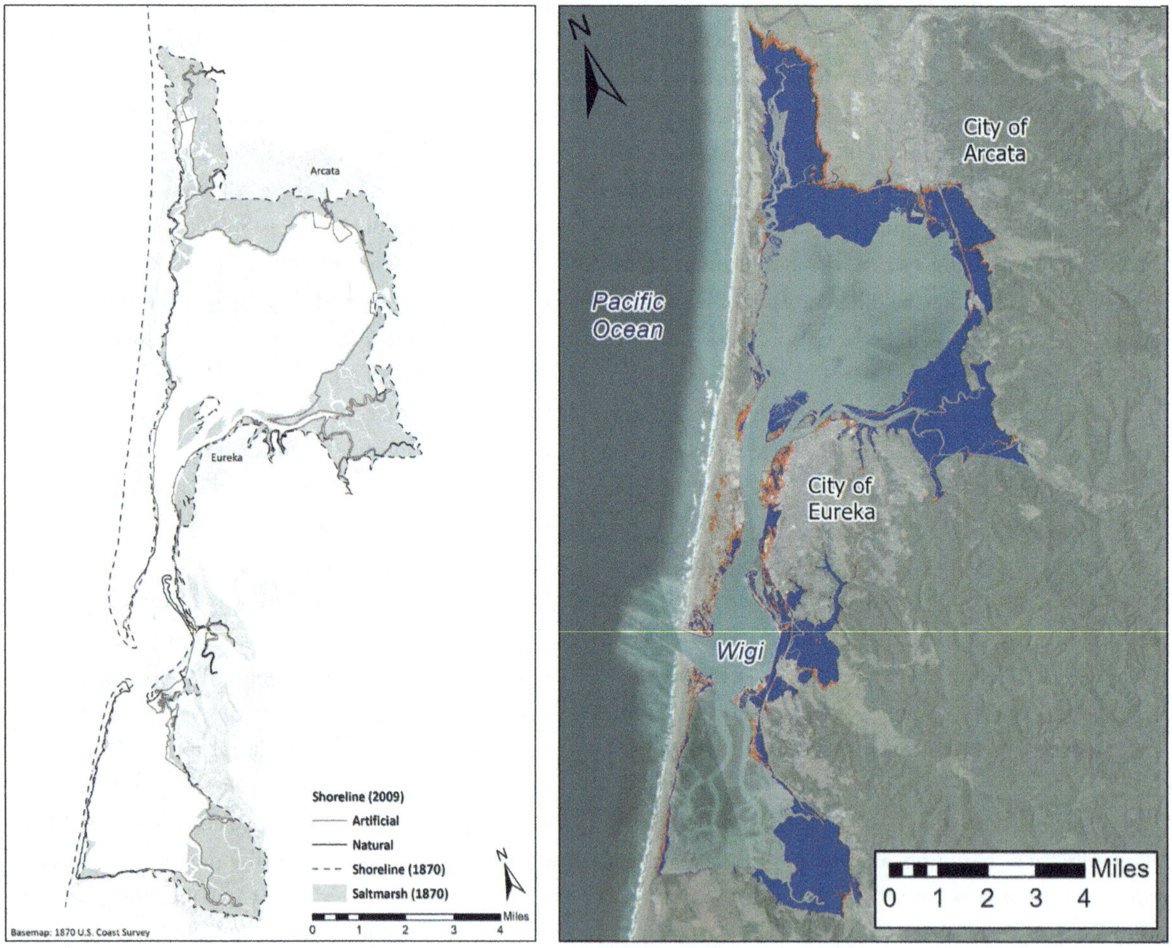

Figure 1: (a) 1870 USGS survey of Wigi, with 1870 shoreline (dotted line) and 2009 shoreline (thin line for artificial and thicker line for natural), serves to illustrate the historic size of the bay (current bay size plus the historic saltmarsh areas in gray) and the magnitude of change to the bay following white settlement. Map: Jay Patton based on data from Laird et al. 2013. (b) Map showing inundation areas with 0.5 m (1.6 feet—projected for 2040) (in the darker blue) and 1.0 meters (3.2 feet—projected for 2065) (in the lighter orange) of relative sea-level rise (Map: Kristen Orth-Gordinier).

Researchers have discovered that at the same time the level of the ocean is rising, the land around Wigi is subsiding, leading to stark SLR projections (NHE 2014; Patton et al. 2017; Figure 1b; Table 1). Local researchers have conducted extensive SLR assessment and planning work which shows that important economic, social, and cultural assets in the Wigi region are at risk from SLR. A recent Humboldt Area Hazard Mitigation Planning effort showed that SLR is likely to affect 2,686 residents and 1,164 buildings whose structure and contents are valued at an estimated $2.3 billion (Tetra Tech 2019). A recent vulnerability assessment revealed that three low-income residential communities along Wigi (King Salmon, Fields Landing, and Fairhaven) are of such low elevation that they have been projected to start experiencing monthly flooding from SLR as early as 2030, and some areas already flood during the highest tides of the year (DWR 2019; Kunkel 2019; Laird 2019).

Table 1. California Ocean Protection Council Projections for relative sea-level rise on Humboldt Bay (Wigi) based on the high-emissions, medium-high risk aversion scenario (OPC 2018).

Year	Projected Rise (ft)	Projected Rise (m)
2030	1 foot	0.3 m
2050	2 feet	0.61 m
2060	3 feet	0.91 m
2100	7.6 feet	2.32 m

Most of the diked former tidelands surrounding Wigi are used for dairy and livestock, field and row crop, and cut flower production, and SLR is likely to reduce their agricultural value as fields become inundated with salt water (Laird 2018b). Crucial transportation and utility infrastructure including Highway 101, regional power generating, and wastewater treatment facilities will be affected (Laird 2018a). Spent nuclear fuel from the decommissioned Pacific Gas & Electric (PG&E) nuclear power plant on Wigi is at risk of being exposed through shoreline/bluff erosion if 2.0 meters of sea level rise is allowed to overtop an existing sea wall (Laird 2019; Brown and Marlow 2023). In 2018, Laird and the Wiyot Tribal Historic Preservation Office identified 52 Wiyot cultural sites likely to be affected by 1.5 meters of SLR (Laird 2018a). Further, SLR is projected to have profound ecological consequences on Wigi as it may shift the composition of coastal wetlands that are important for migratory birds on the Pacific Flyway (Laird 2018a), and rising seas could lead to the flooding of contaminated sites, causing toxins to leach into the bay (Laird 2016).

Strand 1: Indigenous Knowledges and Leadership

The colonial legacy of the United States has contributed to untold social and ecological harms. Genocide and land theft have acted to sever Indigenous connections to, and leadership in, the stewardship of lands within their ancestral territories (Gilio-Whitaker 2020; Fernández-Llamazares et al. 2021). Settler land use and management paradigms radically shifted landscapes, contributing to ecological degradation and setting the stage for the climate crisis (Reibold 2022). SLR provides a transformative opportunity to overcome colonial paradigms for environmental stewardship. This can be done by centering IKs and indigenous leadership in research, planning, implementation related to SLR.

Indigenous Knowledges

IKs are described as inherited wisdoms or bodies of observational knowledge passed down through generations and intricately woven through culture in ways that have allowed Indigenous peoples around the world to sustain themselves in dynamic environments (Cajete 2000; Kimmerer 2000; Bethel et al. 2022). IKs are gaining prominence for their ability to address complex environmental challenges; however, until recently, these ways of knowing were marginalized and often dismissed as mythology or folklore (Lipe 2013). As the original stewards of the land, Indigenous people have coevolved with environments and continued to adapt to changing climates since time immemorial and can contribute great insights to climate resilience.

There are important ethical considerations related to the use of IKs. Scholars warn that it may not be appropriate to separate insights from IKs from their cultural context (Agrawal 2002) and some aspects of IKs may be considered confidential (Kimmerer 2000). Therefore, Tribes and Indigenous knowledge-holders must maintain autonomy over how IKs are collected and shared. Additionally, drawing inspiration from IKs without centering the vision and leadership of the Indigenous Nations from which the knowledge comes, can be considered a form of appropriation (Smith 2021).

Drawing from IKs in climate research and practice contributes to greater understanding of material and physical landscape changes (Hernandez 2022). Further, IKs can contribute to new ethics or value insights, potentially challenging academics and practitioners to transform their worldviews and imagine new relationships with the 'more than human world' (Kimmerer 2013; Johnson and Larsen 2017). Indigenous oral accounts describe the ecological and social history of the Wigi region, including seismic events, geologic processes, ecosystem properties, shifts in land use practices, and movements of people that Western science has only recently been able to articulate—often only partially. IKs can also provide insights into technologies and techniques for adapting to SLR; as an example, the Swinomish Tribe has revived the practice of clam gardening as a technique to prevent coastal erosion and produce culturally important foods (Ryan 2022). The Wiyot Tribe, with support from the SLRI, has received three grants to develop a climate adaptation plan. As a part of this effort, Tribal staff have been conducting interviews with Tribal citizens to gain an understanding of Tribal knowledge and priorities related to SLR and climate change.

Several Indigenous practitioners state that IKs and Western science are not mutually exclusive and that braiding these understandings together can produce a strengthened approach when Indigenous leadership is centered and ethical considerations are addressed (Kimmerer 2013; Lipe 2013; Emery et al. 2014; Matson et al. 2021). Native communities in North America have been taking a leadership role in planning for climate change, often through implementing techniques that draw from Indigenous worldviews along with Western science (Jones 2020). In a genuine practice to braid these understandings in our region, we may be better positioned to approach SLR issues by filling research gaps and bolstering our collective understanding of complex changes facing our region. Local Tribes, including the Wiyot Tribe, have the potential to be innovators in terms of developing SLR resilience strategies that draw from Indigenous knowledges and worldviews. Ecocultural approaches to SLR adaptation might produce ecological and cultural benefits such as supporting habitat for species that provide culturally important foods.

Indigenous Leadership and Land Return

SLR provides the opportunity to elevate Tribes into a leadership role in the planning for the future of coastal resources within their ancestral territories. The SLRI prioritizes Tribal leadership with a Tribal co-chair and a goal to develop projects that are of priority to Tribes. Further work can be done to consider how governance structures related to SLR adaptation and planning around Wigi can codify Tribal leadership and co-management. Cultivating Tribal leadership also requires supporting capacity development among Tribes to build the expertise and personnel to address these key questions. Cal Poly Humboldt can continue to support this process through more research grants and projects that include funding to support Tribal staff time and capacity—building off the momentum from collaborative SLR projects conducted to date with the Wiyot Tribe.

SLR may also provide the opportunity to upend unjust land ownership structures. Returning culturally significant places to the original stewards is a potential SLR resilience strategy that could serve many purposes including (1) addressing historical injustices while bringing healing and empowerment to Tribes, and (2) providing the opportunity for Tribal leadership to implement strategies for adaptation that would benefit all in the region.

Tribal acquisition of coastal properties could accelerate the process to implement SLR and coastal resilience strategies, as upon acquisition, the Tribe could prioritize implementing ecocultural resilience activities. The Wiyot Tribe has already exhibited the capacity to implement such actions. In 2004 when the Tribe acquired portions of Tuluwat Island—a sacred site where toxic waste had been sitting unaddressed for over a decade—they immediately initiated a process to clean up the site which they completed in 2014 (EPA 2018). Without the Tribe's acquisition of the property and leadership on ecocultural restoration, the contaminated site would likely still sit unabated, remaining a pollution risk to all users of the bay, particularly as SLR threatens to inundate the site.

Land return also provides an opportunity for Tribes to expand land ownership as a form of resilience to SLR. For example, the Wiyot Tribe owns less than 1% of the land within their ancestral territory and many of those holdings are within low-lying coastal areas at risk of inundation from SLR. Acquisition of more upland coastal property can help the Tribe to maintain and even grow their landholdings against the threat of land loss from SLR. As an example, the SLRI has partnered with the Wiyot Tribe on a recent OPC funded project that has enabled them to purchase Mouralherwaqh, a 46-acre parcel of coastal property upland of King Salmon, for ecocultural restoration purposes.

Strand 2: Geosciences

Broadly, geoscience is the study of the earth and the processes that shape it, encompassing earth sciences such as geology, oceanography, and atmospheric science. Earth scientists use observations, data analysis, theory and modeling, and other specializations in particular methods (e.g. geospatial scientists who specialize in mapping phenomena on the earth's surface). Geosciences are crucial to understanding the cause, extent and type of impacts from SLR as well as providing a means to map, visualize, and communicate SLR impacts and processes.

Geoscience and SLR

The impact that SLR will have on any given place on the coast is tied to two factors: (1) global SLR or the extent and speed at which seas will rise as a result of climate change and other factors; and (2) the vertical motion of the land in that area or the extent to which the land is either uplifting or subsiding. If the land in a given area is subsiding (going down) at the same time that seas are rising, that can lead to greater SLR effects and conversely, if the land is uplifting, the impact from SLR may be moderated. The rate at which seas are rising in relation to the rate at which land elevation is changing is often referred to as relative sea-level rise (rSLR). Geoscientists can provide insights into both factors that influence rSLR leading to more accurate SLR projections.

Many factors can contribute to vertical land motion (VLM) including subsidence from sediment compaction (Allen 2000), ground water extraction (Ireland, Poland, and Riley 1984), glacial isostatic elevation change (Peltier 1976), and geothermal factors (Massonnet, Holzer, and Vadon 1997). However, the key contributor to VLM in this region is tectonics—namely how the Earth moves in response to interactions between various earthquake faults and tectonic plates. The tectonics of coastal Northern California are dominated by overlapping plate boundaries: the Cascadia subduction zone (CSZ), the San Andreas fault, and the Mendocino fault. The Gorda plate subducts beneath the North America plate at about 36 mm/year to form the CSZ megathrust fault (McCaffrey et al. 2007; Appendix A1). Between large earthquakes, regions directly above the CSZ fault tend to subside and regions landward of the fault tend to uplift (Hyndman and Wang 1995; Wang et al. 2001; Loveless and Meade 2010; Feng et al. 2012).

In the fall of 2010 a group of geoscientists, including individuals connected to the SLRI, formed the Humboldt Bay Vertical Reference System Working Group (HumBayVert) (Patton et al. 2017, 2023) with a goal to evaluate the tectonic contributions to local sea level in Northern California. They used tide gage data, Global Navigation Satellite System data (GNSS; the international version of GPS), and National Geodetic Survey first-order leveling data to help measure land-level change in the southern CSZ region.

This group produced estimates of annual vertical land motion and relative sea-level-rise rates for various sites throughout the California North Coast region (Table 2). Their work shows great variability in VLM across the region; Wigi is subsiding while the surrounding ar-

*Table 2. 20th to 21st Century local relative sea level (rSLR) and vertical land motion (VLM) rates for sites in Northern California.**

Tide Station	Relative Sea Level Rate (mm/yr)	Vertical Land Motion Rate (mm/yr)
Crescent City	-0.84	2.83
Trinidad	2.86	-0.87
Mad River Slough	2.53	-0.54
Samoa	3.92	-1.93
North Spit	5.20	-3.21
Fields Landing	4.65	-2.66
Hookton Slough	6.64	-4.65

*Note: rSLR rates show the rate at which the sea level is changing—negative rates mean the sea level is going down relative to the land, while positive rates mean it is rising relative to the land. VLM rates depict the rate at which the land elevation has been changing—negative rates mean the land is subsiding and positive rates mean it is uplifting. These rates have been developed through looking at past data from tide gages throughout the region and, as such, these data differ from the future projections highlighted in Table 1. Sources: Patton et al. 2023.

eas are not (Table 2; Appendix A2). South of Humboldt Bay there is uplift occurring in the Fortuna-Scotia region, similar to the uplift rates north of Wigi. Rates of VLM are estimated at -4.65 mm/year (subsidence) in Hookton Slough in southern Humboldt Bay, and 2.83 mm/year (uplift) in Crescent City (Table 2). Within the bay, the North Spit and Samoa show the lowest rates of subsidence while Hookton Slough shows the highest rates of subsidence. The rate of subsidence in south Wigi is among the highest on the entire West Coast. Figure 2 in Appendix A depicts VLM rates at the sites used in the Patton et al. (2023) analyses.

Quantifying future local sea-level change is the first step in planning strategies for coastal ecosystems (Church et al. 2011; Horton et al. 2014). Results from the Patton et al. (2023) study provides fundamental sea-level rise data for making management decisions as they apply to coastal landscapes and the species and ecosystems that are the most vulnerable to future sea-level rise (Nicholls and Cazenave 2010; Nicholls 2011). The results from the HumBayVert studies have already been incorporated into sea-level-rise planning in Humboldt County, the cities of Arcata and Eureka, and the communities of Fairhaven, Fields Landing, and King Salmon (Laird 2018). There is a need to build off this work to develop more accurate assessments in the Wigi region and beyond.

Geospatial Analyses

Geographic information science (GIS), spatial analysis, and remote sensing are concerned with the spatial distribution of features on the earth. This includes the movement of water and the resulting impact on our environment and built infrastructure. Cartography is concerned with the creation of maps that can accurately portray spatial information to diverse audiences. These disciplines and their related technologies (such as geographic positioning systems (GPS)) are part of an overall geospatial discipline that is utilized by a wide variety of other disciplines and are key to understanding existing and long-term effects of SLR.

Notable uses of geospatial sciences include work in the Historical Atlas of Humboldt Bay and Eel River Delta, developed by Aldaron Laird in 2007, and the Humboldt Bay Shoreline Inventory, Mapping, and Sea Level Rise Vulnerability Assessment (Laird et al. 2013) which include historical maps back to 1854, aerial photographs from 1948, and maps of the current conditions of the bay. The atlas documents changes in shoreline location and shoreline types around the bay and their varying levels of vulnerability. Updated maps (Anderson 2018) showing the potential areas of future inundation are available on the SLRI website (SLRI 2023).

Overall, SLR may require geoscience disciplines to transform with new types of collaborations across disciplines and sectors, including geoscientists working closely with planners and communities to find more effective ways to communicate findings to a wide audience. Additionally, SLR will require scientists to develop a more precise understanding of VLM and how it is influenced by tectonic processes.

Strand 3: Marine and Coastal Science

Around the world, rSLR in combination with other aspects of climate change is affecting the distribution and ecosystem functions of marine habitats at the same time as it is damaging infrastructure and systems that people depend upon. For example, seagrass beds in both hemispheres support a diverse biota, much of which is commercially and culturally important (Wyllie-Echeverria et al. 2002; Cullen-Unsworth et al. 2014). But as seas rise and beds move upslope to pursue sufficient light, they encounter natural and anthropogenic upland barriers and so become locally extinct when the water becomes too deep (Short and Neckles 1999, Prosser at al. 2018). In this case, ecosystem functions that humans depend upon, like making ocean water less acidic, carbon sequestration, the raising of juvenile salmonids as well as rockfish and crabs, food for birds on the Pacific Flyway, and recreation, are also lost (Duarte 2002, Mazarrasa et al. 2015).

SLR does, however, present an opportunity for a paradigm shift. Local government planning processes understandably focus on infrastructure but usually do not concomitantly include marine habitats (e.g. Humboldt County Civil Grand Jury 2022). A more effective planning process would integrate the responses for infrastructure with concern for vulnerable coastal habitats (e.g., salt marsh, eelgrass, and freshwater marshes). Pragmatic considerations, like the penetration of seawater underneath earthen dikes, and the financial costs of adding new barriers, may necessitate this approach. Marine and coastal scientists may also be able to speak to the potential environmental impacts of various SLR adaptation strategies and assist with the development of more ecologically favorable approaches.

Marine Habitats of Humboldt County

SLR has the potential to dramatically shift the region's marine and coastal habitats, particularly as they interact with artificial shoreline structures installed during the settler period and beyond. Barriers built to hold back estuarine and open coast waters (e.g. earthen dikes) have the effect of blocking access to upland spaces into which marine habitats could shift. For example, the dikes now prevent the upslope migration of salt marsh as sea level rises. With these existing or potentially higher dikes, salt marsh is being replaced by mid intertidal mudflats, which will be replaced by low intertidal eelgrass beds, which in turn will be lost as the water deepens (Figure 2). Similar to the rSLR habitat dynamics within Wigi, if the rate of sand accumulation on the coastal beaches and

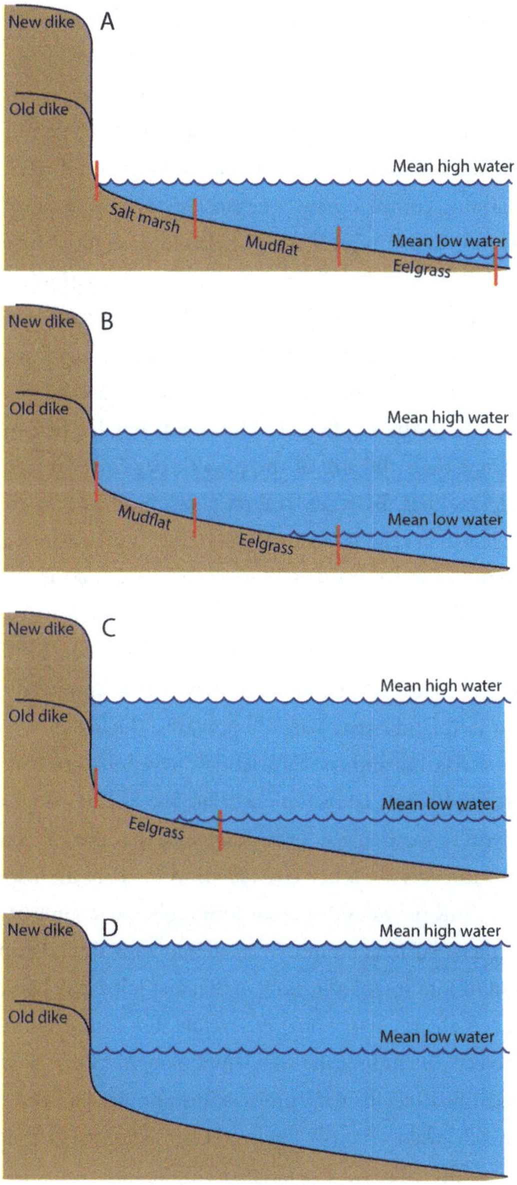

Figure 2: The chronological drowning of marine habitats in Wigi as they encounter an upland barrier during rSLR (illustration by Fiona Shaughnessy).

dunes of Humboldt County does not at least keep pace with global SLR and the downward tectonic movement of the land, then the present locations of these habitats will be overswept.Whether or not they relocate landward will depend on the extent of upland barriers. Most of the migratory and resident birds (e.g., brant goose, western snowy plover), fishes (e.g., coho salmon, black rockfish), invertebrates (e.g., Dungeness crab, clams) and marine mammals (e.g. harbor seals) that have fidelity to one or more of these estuarine or outer coast habitats would experience population crashes.

Although estimates of the global SLR rate continue to increase, the predictions of temporary *increases* of habitat size for some marine habitats in Wigi could result in planning complacency. For example, the spatial extent of eelgrass in southern Wigi is predicted to slightly increase under present day SLR conditions, or dramatically, but temporarily, increase if global SLR rates become greater than 2.8 mm/yr (Shaughnessy et al. 2012). But the only reason that eelgrass area is predicted to increase is because the size of the mudflats in Humboldt Bay happen to be greater than the size of the present day lower intertidal eelgrass beds. In other local estuaries, mudflat area could be less than eelgrass area, in which case future eelgrass area extent would immediately *decrease* in response to rSLR. Regardless of whether rSLR results in the increase or decrease in the extent of any of the intertidal marine habitats over the next several decades, they will all ultimately be lost if upland barriers are too pervasive (Figure 2).

Marine and coastal scientists have much to contribute to planning conversations related to SLR on Wigi and beyond—particularly in making sure that coastal ecosystem impacts are a key consideration in adaptation strategies. SLR provides the opportunity to remove some of the artificial barriers around Wigi to allow for the migration and restoration of eelgrass, mudflat and saltmarsh habitats. This is already happening in several salt marsh restoration projects, including the McDaniel Slough, White Slough, and Elk River Estuary projects, but a more bay-wide approach could seriously forefront ecosystem sustainability.

There are a series of research questions listed in Appendix B that flow directly from this need to more specifically understand how rSLR continues to affect local marine habitats. These questions show that input from a wide range of marine science disciplines—from the physical to the biological—will be necessary for proposing adaptation strategies that maintain marine habitats and ecosystem functions.

Strand 4: Engineering

Engineering will play a crucial role in understanding the potential impacts of climate change and developing effective adaptation strategies.

Understanding and Predicting the Effects of SLR

Engineers, along with scientists, are involved in the prediction of where and how flooding will occur from SLR. Some of the foundational models and assessments related to potential SLR impacts in Wigi were developed by engineers (e.g. Anderson 2015, 2018). In addition to quantifying the expected change in sea level, engineers can estimate the flows of sediment into and out of Wigi, and make predictions about how these flows might change under SLR. This information will inform management strategies to protect the coastal communities around Wigi. For example, Curtis et al. (2021) found that sediment supply to Wigi is projected to increase up to 58% under climate change scenarios, which could provide an opportunity to use marsh ecosystem restoration to enhance coastal protections.

Engineers can also help to understand how SLR will interact with groundwater. Research shows that in many areas, higher sea levels will push up the water table, causing saltwater intrusion into groundwater sources and leading to flooding that bubbles up from underneath the ground (Rotzoll and Fletcher 2013; Pierre-Louis 2021). This means that even if certain areas build up levees or sea walls to keep rising waters out, those areas could still flood from below—calling into question the usefulness of those protection measures. We expect SLR will increase the groundwater level in the areas around Wigi, and groundwater models developed by engineers are essential to understanding these impacts (e.g. Willis 2014). Additional research into aquifer properties would build upon these results to better understand the impacts of SLR in the surrounding groundwater systems.

SLR can affect water quality as well. Numerous low-lying contaminated sites around Wigi are vulnerable to rising sea level and groundwater. Many are former lumber mills that used a wood preservative containing dioxins, while others are former rail yards, boat repair facilities, and other industrial sites contaminated with various pollutants. Rising groundwater can flood these sites from below, pushing contaminants to the surface (May et al. 2020). Mobilizing dioxins, metals, petroleum hydrocarbons, and other industrial contaminants can impact adjacent waterways and expose people as well as wildlife, including commercial, subsistence, and sport fisheries. Engineers can help to identify toxic sites most at risk of inundation and develop effective remediation strategies.

Designing Effective and Resilient Adaptation Strategies

Conventional engineering designs to prevent flooding often focus on hard structures such as sea walls, stream channelization, levees, and riprap. However, these designs often transfer or propagate environmental vulnerabilities (Dugan et al. 2011). Nature-based designs incorporate ecosystem restoration as a strategy to prevent flooding (Pearce 2022). A recent California technical report sought to define and categorize these approaches using the term "natural shoreline infrastructure" defined as "using natural ecological systems or processes to reduce vulnerability to climate change related hazards while increasing the long-term adaptive capacity of coastal areas by perpetuating or restoring ecosystem services" (Newkirk et al. 2018:5). Examples include: vegetated dunes, cobble berms, marsh sills, tidal benches, native oyster reef, and eelgrass beds. Research shows that in addition to having fewer negative environmental impacts, those designs may be more cost effective and provide better flood protection in the long term (Newkirk et al. 2018). Narayan et al. (2016) found that salt marshes can be particularly effective at reducing wave heights, so protecting and expanding these native ecosystems could be an important part of SLR adaptation in Wigi. The majority of protection infrastructure that is currently in place along the coastline rely on conventional hard structures with very few projects that integrate nature-based systems implemented or even studied. This provides an opportunity for Cal Poly Humboldt to play a meaningful role in developing and understanding these new (old) technologies.

Some practitioners have taken to describing SLR protection measures as falling along a spectrum from "gray" to "green," with gray being engineered hard structures and green being entirely nature-based solutions. Hybrid protection measures could fall somewhere in the middle of that spectrum. For example, a levee could be designed with a living shoreline of restored marsh or oyster reef in front of it. Engineers can assess the vulnerabilities of infrastructure to the impacts of SLR and determine what types of infrastructure improvements and protections would be most effective in specific locations. One important example is the stretch of Highway 101 that runs along the east side of Wigi. This is particularly vulnerable to SLR, and Caltrans is considering a number of engineering design solutions that would help to protect this transportation link from frequent flooding (e.g. Burns 2021; Vanderheiden 2022). Caltrans has reached out to the Cal Poly Humboldt Environmental Resources Engineering (ERE) department to develop SLR-related projects, and the fall 2022 capstone projects focused on creating road protection designs that are adaptive to SLR and fish passage. A list of potential areas or topics for continued research can be found in Appendix B.

Complex environmental challenges like SLR have contributed to transformations in fundamental thinking and modes of operating in the field of engineering. Engineering has traditionally been a field that prides itself on expertise and the ability to design and engineer out of problems, like developing sea walls, gates, and levees to prevent flooding. In SLR and flood management, some of that thinking has begun to change. Engineers are questioning hard flood protection measures designed to 'hold the line' and have begun exploring nature-based solutions and other mechanisms that gracefully let water in (Kimmelman and Haner 2017). The field has also been grappling with questions of equity, with a sense that engineers could better work with communities to co-develop design solutions. This has brought a sense of humility to the field and an openness to collaborating across disciplines and ways of knowing. IKs serve as an inspiring example as Tribes have been at the forefront of implementing forms of natural shoreline infrastructure that draw from a cultural perspective (Jones 2020).

Cal Poly Humboldt is positioned to become a leader in community and Indigenous-centered engineering with the launch of the innovative Master's Degree in Engineering & Community Practice which includes a combination of coursework in engineering and Native American studies and has students working on projects with partners from underserved communities including local Tribes. In addition, the Cal Poly Humboldt ERE department has long been a leader in nature-based designs and solutions to complex challenges—most prominently in their work to support the design and creation of the Arcata Wastewater Treatment Facility (WWTF). The unique WWTF, located at the north end of Wigi, relies on a series of constructed wetlands to treat wastewater from Arcata before releasing the cleaned water into the bay. It is listed as one of 17 case studies of wetland treatment systems in the EPA's guidance (EPA 1993).

Strand 5: Social Science, Policy, Law, and Planning

Numerous individuals, communities, and Tribes in the North Coast region will be affected by projected SLR and have a stake in the outcome of SLR planning. This highlights the importance of understanding the socioeconomic context surrounding SLR locally and seeking to develop policy, legal, and planning responses that can help to facilitate equitable, resilient, and transformative SLR adaptation.

Social Science

Scholars emphasize the importance of gaining rich, contextual information about the social conditions in communities facing coastal hazards like SLR in order to understand the potential implications of various strategies and to inform the development of effective and just adaptation actions that respond to the realities being experienced on the ground (Marino 2018b; McAdam 2011; Bettini 2013; Marino 2015; Farbotko and Lazrus 2012; Perumal 2018). Past social science research suggests that SLR adaptation processes are likely to produce inequitable outcomes without anticipatory planning and decision-making that explicitly acknowledges and seeks to overcome inequalities (Siders and Ajibade 2021; Marino 2018a,b; Hardy et al. 2017; Boyer and Penn 2013; Herreros-Cantis et al. 2020).

Sea-level rise provides a transformative opportunity for researchers to develop social science in coordination with communities and local governments that directly speaks to local SLR planning processes. Relevant research areas include: exploring the sociocultural context of coastal communities and their sense of place; documenting local knowledge, values, and priorities related to SLR; and understanding social dynamics related to coastal governance. Various researchers from Cal Poly Humboldt have engaged these issues by pursuing collaborative research projects with community partners, several of which have directly contributed to local government processes (e.g. Blakeney et al. 2011; Kunkel 2019; Carrasco et al. 2021; Orth-Gordinier 2022; Brown and Marlow 2023).

Planning

The field of coastal planning has been undergoing transformations over the past decade in response to the challenges from increased coastal hazards and SLR, as well as calls for more inclusive and just approaches (Swanson 2021; APA 2023). In the past, planners have approached challenges from coastal flooding by increasing protection measures such as dikes and levees and ensuring that home and business owners had access to flood insurance to protect those assets. While the goal of these strategies was to increase safety, paradoxically, these efforts made those areas less safe as they allowed for the development of more infrastructure in risky areas, leading to more devastation following flood and hazard events (Burby 2006). As a result, planners have moved toward considering approaches that prevent development in flood prone areas and incentivize the relocation of infrastructure and homes from flood vulnerable areas through a process sometimes referred to as "strategic retreat" (Flavelle 2020). Strategic or managed retreat programs mark a turning point, orienting away from reactive programs (such as the National Flood Insurance Program (NFIP), which provides payouts to home and business owners to rebuild their homes after a flood has occurred) toward encouraging more proactive approaches to SLR and coastal flood management. The outcomes of managed retreat programs vary widely, depending on their approaches. Efforts that aim to solely minimize physical risk (e.g., a "techno-managerial" approach), are less likely to center justice at the various scales needed to produce structural change and "transformative" outcomes (Ajibade et al. 2022).

Local governments and Tribes in the Wigi region have begun efforts to consider and plan for future SLR (see the SLRI Library Database for links to local reports and plans), but there remains a need to expand and coordinate these planning efforts locally and in the North Coast region. Planners will need to consider an array of adaptation strategies, such as developing sea walls or living shorelines; designing accommodation measures such as causeways; or abandoning inundation zones to retreat inland. Complex and potentially conflict-engendering decisions will need to be made about which areas to protect (and how) and which areas to retreat from (and when). The term 'retreat' has been controversial in SLR discussions in many parts of the world, as residents with attachments to the places where they live may interpret "retreat" as a form of "giving up" (Anderson 2022). Equity considerations arise as well, since basing decisions about retreat solely on economic factors (i.e. protecting places of the highest real estate value and retreating from others) can serve to further marginalize low income communities and exacerbate inequalities (Marino 2018a,b; Siders and Ajibade 2021).

SLR adaptation measures have the opportunity to reshape development and will have significant implications for all connected to the North Coast region. The work and practice involved with ensuring that decision-making processes are inclusive and equitable will require meaningful efforts to educate and engage the community on the issue. Such tasks will require strategic investment and innovation in community outreach and engagement, beginning with Indigenous communities. Wigi can gain inspiration from long-term water education initiatives such as the Water Leaders Institute (WLI) in New Orleans that offer creative and art-centered approaches to community learning (WLI 2022). For these reasons, some planners have argued that 40-50% of budgets for climate adaptation projects should go toward "environmental education and community-based planning processes" (Chang 2018).

Law and Policy

Climate change and SLR present novel challenges to conventional policy and legal approaches to environmental change. SLR sits precariously at the nexus of local, state, and federal authorities, with limited federal and international frameworks to guide a coordinated approach. Under the California Coastal Act, for example, SLR planning and implementation authority is shared by local and state governments, with Tribal governments in a "consultation role" (see, e.g. California Coastal Commission 2018). City and county governments prepare local coastal programs, which, upon approval by the California Coastal Commission (CCC), guide the issuance of Coastal Development Permits (CDP) in the region. However, the CCC retains permitting authority and the ability to hear appeals over certain coastal lands. The CCC retains jurisdiction over a large amount of land on the Wigi shores, which can lead to planning paralysis in SLR vulnerable areas, as neither the local government nor the state entities nor the Tribal governments are certain who should lead the effort (Laird 2018c). This complex regulatory framework evokes traditional environmental law tensions over how local and Tribal governments—those with specific knowledge and expertise over their territories, not to mention sovereignty—exercise agency over localized yet transboundary problems such as SLR (e.g. Fox 2020).

Emerging legal research is examining such tensions and asking novel questions, such as how to characterize the public trust doctrine's jurisdiction over tidelands as they migrate landward due to sea level rise, thus altering the boundaries formerly used to demarcate where public lands end and private lands begin (Lester 2021). The interaction between SLR adaptation processes and traditionally governing environmental laws and regulations remains a critical topic for the evolution of law in a climate-changed world. Other legal questions engaging this tension include how to streamline permitting processes to ensure that SLR mitigation projects can be completed on time to provide flood and other protections, while not subordinating robust public engagement and collaborative planning—as required by the National Environmental Policy Act and the California Environmental Quality Act—to urgent timelines (e.g., BRRIT 2022; CNRA 2022; Ajibade 2022).

The state of California is turning toward these questions as a national leader in establishing a bold climate policy agenda that frames climate change as a present issue rather than a distant future (e.g., the $54 billion climate action plan in Newsom 2022). California has also become an incubator of SLR policy and legislation. State adaptation strategies and policy proposals are beginning to prioritize investment in low-income communities, Tribal communities, communities of color, and communities most marginalized by poverty and climate impacts. The California Ocean Protection Council's Strategic Plan (OPC 2020), Equity Plan (OPC 2022), Tribal Engagement Strategy (OPC 2023), and SLR Guidance reports (OPC 2018) each include recommendations related to equitable SLR adaptation. For example, OPC's Equity Plan states that one of its goals is to "[a]dvance equity across ocean and coastal policies and actions" (see OPC 2022, Goal 2:19). California has also implemented Governor Gavin Newsom's Nature Based Solutions Executive Order N-82-20 (2020), which commits the state to the goal of conserving 30 percent of lands and coastal waters by 2030 (30x30). A specific initiative of 30x30 proposals in California is to provide $100 million in funding to help "implement tribal priorities including ancestral land access, co-management, conservation and return of tribal lands" (Newsom 2022).

In 2021 alone, the California state legislature considered over 12 bills directly related to SLR (California Coastal Commission 2021), including Senate Bill 1078, which proposed a novel strategy for proactive managed retreat through the creation of a Sea Level Rise Revolving Loan Pilot Program providing loans to local governments to support voluntary retreat efforts (the bill was vetoed by Governor Newsom September 29, 2022). Assembly Bill 1384, the Resiliency Through Adaptation, Economic Vitality, and Equity Act, adopted in September 2022, directs the Natural Resource Agency to work with the Office of Planning and Research (OPR) to identify vulnerabilities to communities disproportionately impacted by climate change, and coordinate state budgets to address these impacts, among other responsibilities. Whether transformative structural change will emanate from California's recent legislative and policy initiatives is an open question, but recent actions at least signal that the state seeks forward momentum in this direction.

The Cal Poly Humboldt SLRI can play a leading role in tracking California SLR law and policy development and assessing its effectiveness at facilitating equitable, resilient, and transformative change. The SLRI can provide recommendations and guidance for how existing and future laws and regulations can fund and support integrated and just SLR adaptation that enhances the resilience of the North Coast region and beyond.

Discussion

Sea-level rise transcends all kinds of boundaries—political, disciplinary, organizational, ecological, and ways of knowing. It can only begin to be understood and addressed if different disciplines and types of knowledge are brought together. Each approach and discipline brings an important lens to the table. Even as we were writing individual disciplinary sections in this paper, we found it difficult as lines between the approaches were blurry and uncertain. For example, the questions related to the hydrologic implications of SLR can be explored by engineers, geologists, oceanographers and watershed scientists. Considerations around the design of natural shoreline infrastructure draw from engineering, but also IKs and marine and coastal ecosystem science as many of those designs involve restoring coastal ecosystem habitats. Robust planning and policy incorporates knowledge and information from all fields and perspectives.

In this context, Indigenous worldviews can continue to be a source of inspiration. Lipe (2013) describes IKs as 'high context' ways of knowing, where "knowledge is gained through the understanding of many different variables at many different levels simultaneously" (Cajete 2008). With Kimmerer's analogy of the braid, we can envision these different approaches or strands winding together and, in the winding, shaping and influencing one another until neither the individual disciplines nor the overall approach remain the same. Also, braided or woven materials are much stronger than individual strands.

The SLRI was founded to provide a space for braiding these different ways of knowing. To do this we have sought to create a network that transcends disciplinary and organizational boundaries. The governance structure requires two co-chairs, one who approaches SLR issues from Cal Poly Humboldt's academic sphere, and the other from a Tribal or community position. In building SLR projects and research, we seek to listen to and address Tribal and community needs, rather than solely focusing on theoretical concepts of interest to specific disciplines. We are flexible to develop leadership and funding structures most suitable to individual projects. For many projects, Tribal, community, or local government entities are suited to lead the effort while Cal Poly Humboldt plays a supporting role. In some projects, it may make sense for private consulting firms to play a large role in developing project outputs. A strong network allows all the participants to call upon one another as expertise and resources are needed. We seek to ensure that projects are not duplicative and build upon one another. We also strive to keep students at the center of the network's operation and activities to provide learning opportunities and to gain inspiration from students' insights and perspectives.

Further developing a focus on climate and SLR resilience will help Cal Poly Humboldt achieve its vision as a "comprehensive polytechnic university for the 21st-century" that meets "workforce needs of our region and state" (HSU 2021). This will require developing disciplinary expertise as well as prioritizing avenues for bringing diverse perspectives together. SLRI members have identified the need to prioritize adding faculty expertise related to engineering, geoscience, and marine and coastal ecology aspects of SLR. In collaboration with Tribes and the Native American Studies Department, the institution can further develop expertise and educational programs related to IKs and food sovereignty and integrate these concepts into teaching and research. This could also include developing residency programs to sponsor Indigenous and local knowledge holders to bring their visions to campus.

SLRI faculty have already played a leading role in integrating SLR concepts into curriculum across the campus, including sponsoring capstone projects in Environmental Science and Management, Engineering, and Geospatial Science, where students explore and address SLR issues of interest to the local community. The SLRI includes funded positions for student researchers to engage in SLR activities. Going forward, Cal Poly Humboldt could expand upon these efforts, providing more opportunities for students, particularly underrepresented and Indigenous students, to gain training related to SLR. Some of these opportunities could be facilitated through collaborations with the Indian Natural Resources, Science, and Engineering Program (INRSEP+) and Indian Teacher & Educational Personnel Program (ITEPP). Student engagement in SLR research and education will prepare them to enter the rapidly expanding climate and SLR resilience workforce.

Cal Poly Humboldt will also need to focus on developing research and teaching structures that allow for cross-disciplinary and cross-sectoral collaboration. This could include creating the space to offer SLR and climate resilience classes co-taught by individuals of different disciplines or by a combination of faculty, community, and Tribal experts. This could also include developing more collaborative research spaces that would allow engineers, scientists, IK practitioners, and community planners to work together, perhaps through the development of indoor and/or outdoor research sites that are in the community and adjacent to Wigi. Emerging master's programs in engineering & community practice, marine science, along with discussions to develop a climate resilience master's program expand the opportunities for graduate students to develop integrative thinking and innovate the field even further. Currently SLR work at Cal Poly Humboldt is concentrated in the College of Natural Resources. However, moving forward, there is a possibility for a more expansive vision across the colleges that also incorporates the liberal arts. Fields such as political science, economics, environmental studies, Native American studies, art, literature, journalism, education, history, critical race and gender studies, social work, and other academic centers on campus have a potential role to play in an expansive vision for the topic.

The SLRI has many ideas for future work to move their vision forward. One such idea would be to support the development of a pilot natural shoreline infrastructure project for Wigi as a means to test and demonstrate the efficacy of equitable, resilient, and transformative SLR adaptation solutions. Tribal knowledge and leadership could be centered in the project by incorporating IKs, supporting Tribal leadership in planning and, if possible, by pursuing implementation on Tribally-owned or acquired sites. Engineering, geosciences, marine and coastal science, environmental science and policy, Native American studies, and IKs can all contribute to the design of an infrastructure project that would be effective for SLR protection and ecoculturally beneficial. The infrastructure planning will also need to hold up principles of environmental justice as critical design criteria (e.g. de Schipper et al. 2020). Planning and policy practitioners can assist with community involvement in project planning and implementation, developing government and private industry partnerships, navigating permitting processes, and identifying equitable funding mechanisms. Such a project would require the development of strong partnerships. A project such as this could serve as a demonstration site and educational opportunity. Students could be involved in long-term management, monitoring, and adaptive management of the project.

The emerging work of the SLRI can provide a foundation for developing a vision for transformative climate research and practice. A recent example is the Reclaiming Mouralherwaqh project where the Wiyot Tribe applied for and received a $1.2 million grant from the OPC to acquire a culturally important parcel of land and develop an ecocultural restoration plan for the site. The Tribe had prioritized acquisition of forested lands within the coastal zone and identified Mouralherwaqh as a target property for acquisition. The site contains numerous culturally important plants—many of which are not found on any of the Tribe's current landholdings. Several plants are the types used in basketweaving and ceremonial regalia. The SLRI provided support for that effort when needed, following the directive and leadership from the Tribe. The parcel was acquired in July of 2022 and the Tribe now plans to develop an ecocultural restoration that incorporates cultural use, food sovereignty, water quality, watershed/forest restoration, SLR resilience, law and policy, and other areas of interest to the Wiyot Tribe. There is funding to incorporate Cal Poly Humboldt students and faculty across several different departments into the site planning.

In August 2022, the Wiyot Tribe hosted a Mouralherwaqh Coming Home Ceremony. The Ceremony included Wiyot dances; speeches from Tribal, academic, and government partners; and a shared meal of traditional foods. As academic, community, and government representatives were there with Tribal citizens and staff to witness and celebrate the Tribe's achievement, it was possible to imagine a new paradigm for academic research and education that overcomes colonial power structures. Instead of seeking to study Tribes and communities, researchers can work to support them in achieving their priorities. In the words of Hilanea Wilkinson, a young citizen of the Wiyot Tribe, this pathway can be possible if we learn to transform and "work together."

Acknowledgements

The work and analysis in this paper stands on the shoulders of an entire community of folks who have cared for and stewarded California North Coast coastal resources from time immemorial to today. Understanding and addressing SLR has and will continue to be a whole community effort. Cal Poly Humboldt is located on the present and ancestral Homeland and unceded territory of the Wiyot people. The land that Arcata rests on is known in the Wiyot language as Goudi'ni. Past actions by local, state, and federal governments removed the Wiyot from this land and threatened to destroy them. We acknowledge the Wiyot community, their elders both past and present, as well as future generations as a statement of commitment to actions that dismantle the ongoing legacies of settler colonialism. Funding to support the organization, writing, and editing of this collaborative work was provided by the Cascadia Coastlines and Peoples Hazards Research Hub, which is supported by the National Science Foundation award NSF-2103713. The preparation of this publication was also supported by NOAA Grant # NA22OAR4170106, through NOAA's National Sea Grant College Program, U.S. Dept. of Commerce. The statements, findings, conclusions and recommendations expressed in this paper are those of the authors and do not necessarily reflect the views of the agencies or entities for whom they work or by whom they are funded.

References

Agrawal, Arun. 2002. "Indigenous Knowledge and the Politics of Classification." *International Social Science Journal* 54(173):287–97.

Ajibade, Idowu, Meghan Sullivan, Chris Lower, Lizzie Yarina, and Allie Reilly. 2022. "Are Managed Retreat Programs Successful and Just? A Global Mapping of Success Typologies, Justice Dimensions, and Trade-Offs." *Global Environmental Change* 76:102576. doi:10.1016/j.gloenvcha.2022.102576

Allen, J. R. L. 2000. "Morphodynamics of Holocene salt marshes: a review sketch from the Atlantic and Southern North Sea coasts of Europe." *Quaternary Science Reviews* 19:1155-1231.

Anderson, Jeff. K. 2015. "Humboldt Bay: Sea Level Rise, Hydrodynamic Modeling, and Inundation Vulnerability Mapping: Final Report." Prepared for State Coastal Conservancy, and Coastal Ecosystems Institute of Northern California.

Anderson, Jeff K. 2018. "Sea Level Rise in the Humboldt Bay Region – Update 1." https://digitalcommons.humboldt.edu/cgi/viewcontent.cgi?article=1019&context=hsuslri_local.

Anderson, Ryan B. 2022. "The taboo of retreat: The politics of sea level rise, managed retreat, and coastal property values in California." *Economic Anthropology* 9(2):284-296.

APA (American Planning Association). 2023. "Our Equity, Diversity, and Inclusion Commitment." American Planning Association. Retrieved February 10, 2023. https://www.planning.org/diversity/goals/.

Barnhardt, Ray, and Angayuqaq Oscar Kawagley. 2005. "Indigenous knowledge systems and Alaska native ways of knowing: Indigenous Epistemologies and Education - Self-Determination, Anthropology, and Human Rights." *Anthropology & Education Quarterly* 36(1):8–23.

Barnhart, Roger A., Milton J. Boyd, and John E. Pequegnat. 1992. *The Ecology of Humboldt Bay, California: An Estuarine Profile (Biological Report 1)*. Washington, DC: US Department of the Interior, Fish and Wildlife Service.

Bateson, Gregory. 1979. *Mind and Nature: A Necessary Unity*. 1st ed. New York: Dutton.

Beatley, Timothy. 2009. *Planning for Coastal Resilience: Best Practices for Calamitous Times*. Washington, DC: Island Press.

Bennett, Nathan J., Laure Katz, Whitney Yadao-Evans, Gabby N. Ahmadia, Scott Atkinson, Natalie C. Ban, Neil M. Dawson, Asha de Vos, Juno Fitzpatrick, David Gill, Mael Imirizaldu, Naia Lewis, Sangeeta Mangubhai, Leah Meth, Ella-Kari Muhl, David Obura, Ana K. Spalding, Angelo Villagomez, Daniel Wagner, Alan White, and Aulani Wilhelm. 2021. "Advancing Social Equity in and Through Marine Conservation." *Frontiers in Marine Science* 8:1-13.

Berkes, Fikret. 1999. *Sacred Ecology: Traditional Ecological Knowledge and Resource Management*. Philadelphia, PA: Taylor & Francis.

Berkes, Fikret, Johan Colding, and Carl Folke. 2000. "Rediscovery of Traditional Ecological Knowledge as Adaptive Management." *Ecological Applications* 10(5):1251–62.

Bethel, Matthew B., DeWitt H. Braud, Tara Lambeth, Donald S. Dardar, and Patty Ferguson-Bohnee. 2022. "Mapping risk factors to climate change impacts using traditional ecological knowledge to support adaptation planning with a Native American Tribe in Louisiana." *Journal of Environmental Management* 301:113801. doi:10.1016/j.jenvman.2021.113801.

Bettini, Giovanni. 2013. "Climate barbarians at the gate? A critique of apocalyptic narratives on 'climate refugees'." *Geoforum* 45:63-72.

Bibby, Kiya, and Hayley Carter. 2022. *Toward More Equitable Nature-Based Coastal Adaptation in California: Recommendations for Improving Social Equity in Funding, Policy, and Research.* California Ocean Science Trust. Retrieved August 2, 2022. https://www.oceansciencetrust.org/projects/equitylivingshorelines/.

Blakeney, Victoria, Wade Bonney, Nicholas Chang, Leah Healey, and Elmer. 2011. "Implications of Sea Level Rise on North Humboldt Bay." Senior Planning Practicum Project, Department of Environmental Science and Management, Humboldt State University.

Boyer, Michael, and Erica Penn, 2013. "Tidal Turmoil: Environmental Justice and Sea Level Rise in Hampton Roads: Norfolk Case Study." Virginia Coastal Policy Center 11. Retrieved February 5, 2023 https://scholarship.law.wm.edu/vcpclinic/11.

Brown, Alec, and Jennifer Marlow 2023. *44 Feet Project.* Retrieved February 3, 2023. https://www.44feetproject.com/.

BRRIT (San Francisco Bay Restoration Regulatory Integration Team). 2022. Retrieved September 7, 2022. https://www.sfbayrestore.org/san-francisco-bay-restoration-regulatory-integration-team-brrit.

Burby, Raymond J. 2006. "Hurricane Katrina and the paradoxes of government disaster policy: Bringing about wise governmental decisions for hazardous areas". *The Annals of the American Academy of Political and Social Science,* 604(1):171-191.

Burns, Ryan. 2021. "A Causeway Over the Bay Between Arcata and Eureka? Caltrans Exploring Adaptations for Sea Level Rise." *Lost Coast Outpost.* January 15. Retrieved July 15, 2022 https://lostcoastoutpost.com/2021/jan/15/causeway-over-bay-between-arcata-and-eureka-caltra/.

Cajete, Gregory. 2000. *Native Science: Natural Laws of Interdependence.* Santa Fe, New Mexico: Clear Light Publishers.

Cajete, Gregory. 2008. "Seven orientations for the development of indigenous science education." *Handbook of Critical and Indigenous Methodologies.*, edited by N. K. Denzin, Y. S. Lincoln, and L. T. Smith. Los Angeles: Sage.

California Coastal Commission. 2018. *California Coastal Commission Adopted Tribal Consultation Policy.* Adopted August 8, 2018. https://documents.coastal.ca.gov/assets/env-justice/tribal-consultation/Adopted-Tribal-Consultation-Policy.pdf.

California Coastal Commission. 2021. From John Ainsworth and Sarah Christie to Coastal Commission and Interested Persons. April 14, 2021. https://documents.coastal.ca.gov/reports/2021/4/W6c/W6c-4-2021.pdf.

California Coastal Commission. 2022. "Draft Public Trust Guiding Principles and Action Plan (June 2022)" Issued May 26, 2022. https://documents.coastal.ca.gov/reports/2022/6/Th6e/Th6e-6-22-exhibits.pdf.

California Sea Grant. 2023. "California Sea Grant 2024-2027 Strategic Plan." https://caseagrant.ucsd.edu/sites/default/files/2024-27%20CA%20Sea%20Grant%20Strategic%20Plan_Web.pdf.

Carrasco, Jackson, Jessica Clark, Kelsey Fletterick, Natalie Greenleaf, and Bente Jansen. 2021. "Equitably Engaging Communities Affected by Sea-Level Rise in the Planning Process." Senior Planning Practicum Project, Department of Environmental Science and Management, Humboldt State University.

Chang, Aron. 2018. "Cities are planning for climate change all wrong." *Washington Post.* October 22. Retrieved August 1, 2022. https://www.washingtonpost.com/news/theworldpost/wp/2018/10/22/climate-change-2/.

Cheng, Antony S. 2006. "Build It and They Will Come? Mandating Collaboration in Public Lands Planning and Management." *Natural Resources Journal* 46(4):841–858.

Church, John A., Jonathan M. Gregory, Neil J. White, Skye M. Platten, and Jerry X. Mitrovica. 2011. "Understanding and Projecting Sea Level Change." *Oceanography* 24(2):130-143.

Church, J.A., P.U. Clark, A. Cazenave, J.M. Gregory, S. Jevrejeva, A. Levermann, M.A. Merrifield, G.A. Milne, R.S. Nerem, P.D. Nunn, A.J. Payne, W.T. Pfeffer, D. Stammer and A.S. Unnikrishnan. 2013. "Sea Level Change." *Climate Change 2013: The Physical Science Basis. Contribution of Working Group I to the Fifth Assessment Report of the Intergovernmental Panel on Climate Change.* . Cambridge, UK and New York, NY: Cambridge University Press.

CIBA (California Indian Basketweavers Association) 2022. "Our Vision." Retrieved September 2022. https://ciba.org/our-vision/.

Cullen-Unsworth, Leanne C., Lina Mtwana Nordlund, Jessica Paddock, Susan Baker, Len J. McKenzie, and Richard KF Unsworth. 2014. "Seagrass Meadows Globally as a Coupled Social–Ecological System: Implications for Human Wellbeing." *Marine Pollution Bulletin* 83(2):387–97.

Curtis, Jennifer A., Lorraine E. Flint, Michelle A. Stern, Jack Lewis, and Randy D. Klein. 2021. "Amplified Impact of Climate Change on Fine-Sediment Delivery to a Subsiding Coast, Humboldt Bay, California." *Estuaries and Coasts* 44(8):2173–93.

de Schipper, Matthieu A., Bonnie C. Ludka, Britt Raubenheimer, Arjen P. Luijendijk, and Thomas A. Schlacher. 2021. "Beach Nourishment Has Complex Implications for the Future of Sandy Shores." *Nature Reviews Earth & Environment* 2(1):70–84.

Duarte, Carlos M. 2002. "The Future of Seagrass Meadows." *Environmental Conservation* 29(2):192–206.

Dugan, Josh, Laura Airoldi, Gee Chapman, Simon Walker, and Thomas Schlacher. 2011. "Estuarine and Coastal Structures: Environmental Effects, a Focus on Shore and Nearshore Structures." Pp. 17–41 in *Treatise on Estuarine and Coastal Science*. Vol. 8, *Human-Induced Problems (Uses and Abuses)*, edited by M. J. Kennish, and M. Elliott. Amsterdam: Elsevier.

DWR (California Department of Water Resources). 2019. "Disadvantaged Communities Mapping Tool." Retrieved August 5, 2022. https://gis.water.ca.gov/app/dacs/.

Emery, Marla R., Alexandra Wrobel, Mark H. Hansen, Michael Dockry, W. Keith Moser, Kekek Jason Stark, and Jonathan H. Gilbert. 2014. "Using Traditional Ecological Knowledge as a Basis for Targeted Forest Inventories: Paper Birch (*Betula Papyrifera*) in the US Great Lakes Region." *Journal of Forestry* 112(2):207–14.

EPA (Environmental Protection Agency). 1993. *Constructed Wetlands for Wastewater Treatment and Wildlife Habitat: 17 Case Studies*. EPA832-R-93-005. September 1993. https://www.epa.gov/wetlands/constructed-wetlands-wastewater-treatment-and-wildlife-habitat-17-case-studies.

EPA (Environmental Protection Agency, Region 9). 2018. *The Tuluwat Village Site on Indian Island in Humboldt County, California*. Environmental Stewardship and Cultural Preservation on California's Coast. *Case Study*.

Farbotko, Carol, and Heather Lazrus. 2012. "The first climate refugees? Contesting global narratives of climate change in Tuvalu." *Global Environmental Change* 22:382-390.

Fernández-Llamazares, Álvaro, Dana Lepofsky, Ken Lertzman, Chelsey Geralda Armstrong, Eduardo S. Brondizio, Michael C. Gavin, Phil O'B Lyver, George P. Nicholas, Pua'ala Pascua, Nicholas J. Reo, Victoria Reyes-García, Nancy J. Turner, Johanna Yletyinen, E. N. Anderson, William Balée, Joji Cariño, Dominique M. David-Chavez, Christopher P. Dunn, Stephen C. Garnett, Spencer Greening (La'goot), Shain Jackson (Niniwum Selapem), Harriet Kuhnlein, Zsolt Molnár, Guillaume Odonne, Gunn-Britt Retter, William J. Ripple, László Sáfián, Abolfazl Sharifian Bahraman, Miquel Torrents-Ticó, and Mehana Blaich Vaughan. 2021. "Scientists' Warning to Humanity on Threats to Indigenous and Local Knowledge Systems." *Journal of Ethnobiology* 41(2):144–69.

Feng, Lujia, Andrew V. Newman, Marino Protti, Víctor González, Yan Jiang, and Timothy H. Dixon. 2012. "Active Deformation near the Nicoya Peninsula, Northwestern Costa Rica, between 1996 and 2010:

Interseismic Megathrust Coupling." *Journal of Geophysical Research: Solid Earth* 117(B6).

Flavelle, Christopher. 2020. "U.S. Flood Strategy Shifts to 'Unavoidable' Relocation of Entire Neighborhoods". *New York Times*. August 26. Retrieved August 1, 2022 https://www.nytimes.com/2020/08/26/climate/flooding-relocation-managed-retreat.html.

Fox, Sarah. 2020. "Localizing Environmental Federalism". *UC Davis Law Review* 54:133–94.

Frederikse, Thomas, Felix Landerer, Lambert Caron, Surendra Adhikari, David Parkes, Vincent Humphrey, Sönke Dangendorf, Peter Hogarth, Laure Zanna, Lijing Cheng, and Yun-Chao Wu. 2020. "The Causes of Sea-Level Rise since 1900". *Nature* 584(7821):393–97.

Gaziulusoy, A. Idil, Chris Ryan, Stephen McGrail, Philippa Chandler, and Paul Twomey. 2016. "Identifying and Addressing Challenges Faced by Transdisciplinary Research Teams in Climate Change Research." *Journal of Cleaner Production*, 123:55–64.

Gilio-Whitaker, Dina. 2019. *As Long as Grass Grows: The Indigenous Fight for Environmental Justice from Colonization to Standing Rock*. Boston: Beacon Press.

Hardy, Dean, Richard Milligan, and Nik Heynen. 2017. "Racial coastal formation: The environmental injustice of colorblind adaptation planning for sea-level rise." *Geoforum* 87:62-72.

Hernandez, Jessica. 2022. *Fresh Banana Leaves: Healing Indigenous Landscapes Through Indigenous Science*. New York: North Atlantic Books.

Herreros-Cantis, Pablo, Veronica Olivotto, Zbigniew J. Grabowski, and Timon McPhearson. 2020. "Shifting Landscapes of Coastal Flood Risk: Environmental (in)Justice of Urban Change, Sea Level Rise, and Differential Vulnerability in New York City." *Urban Transformations* 2(1):1-28.

Horton, Benjamin P., Stefan Rahmstorf, Simon E. Engelhart, and Andrew C. Kemp. 2014. "Expert Assessment of Sea-Level Rise by AD 2100 and AD 2300." *Quaternary Science Reviews* 84:1–6.

Humboldt County. 2018. "Sea Level Rise Policy Background Study." July 2018. https://humboldtgov.org/documentcenter/view/64596.

Humboldt County. 2022. "Humboldt Regional Climate Action Plan: Public Review Draft." April 7, 2022. https://humboldtgov.org/DocumentCenter/View/106404/Humboldt-Regional-CAP----Public-Review-Draft-4-7-22-PDF.

Humboldt County Civil Grand Jury. 2022. "The Sea Also Rises." Retrieved on June 14, 2022. https://digitalcommons.humboldt.edu/hsuslri_local/41/.

Humboldt State University. 2021. "A Polytechnic Prospectus." *Humboldt State University*. Arcata, CA.

Hyndman, R. D., and K. Wang. 1995. "The rupture zone of Cascadia great earthquakes from current deformation and the thermal regime". *Journal of Geophysical Research* 100(22): 122-154.

Ireland, R. L., Joseph F. Poland, and Francis S. Riley 1984. "Land Subsidence in San Joaquin Valley, California, as of 1980." *U.S. Geological Survey*, 437-1.

Jevrejeva, S., A. Grinsted, and J. C. Moore. 2009. "Anthropogenic Forcing Dominates Sea Level Rise since 1850." *Geophysical Research Letters* 36(20).

Johnson, J. T., and S.C. Larsen. 2017. *Being Together in Place Indigenous Coexistence in a More Than Human World*. University of Minnesota Press.

Johnson, Martha. [1992] 1998. *Lore: Capturing Traditional Environmental Knowledge*. Hay River, NWT, Canada: Dene Cultural Institute and the International Development Research Centre. Reprint, Darby PA: Diane Publishing.

Jones, Nicola. 2020. "How native tribes are taking the lead on planning for climate change". *Yale E360*, February 11. Retrieved July 12, 2022 https://e360.yale.edu/features/how-native-tribes-are-taking-the-lead-on-planning-for-climate-change.

Kimmelman, Michael, and Josh Haner. 2017. "The Dutch Have Solutions to Rising Seas. The World is Watching." *New York Times*, June 15. Retrieved July 17, 2022. https://www.nytimes.com/interactive/2017/06/15/world/europe/climate-change-rotterdam.html.

Kimmerer, Robin Wall. 2000. "Native knowledge for native ecosystems". *Journal of Forestry* 98(8):4-9.

Kimmerer, Robin Wall. 2013. *Braiding Sweetgrass: Indigenous Wisdom, Scientific Knowledge and the Teachings of Plants*. Minneapolis, MN: Milkweed Editions.

Klein, Julie T. 2008. "Evaluation of Interdisciplinary and Transdisciplinary Research: A Literature Review." *American Journal of Preventive Medicine 35*(2, Supplement):S116–23.

Kunkel, Kristina. 2019. "Exploring Community Knowledge and Perceptions of Flooding and Sea-level Rise in King Salmon, California." Master's Thesis, Department of Environmental Science and Management, Humboldt State University.

Laird, Aldaron. 2016. "City of Eureka Sea Level Rise Assets Vulnerability and Risk Assessment." *Humboldt Digital Commons*. June 2016. https://digitalcommons.humboldt.edu/hsuslri_local/9/.

Laird, Aldaron. 2018a. "City of Arcata Sea Level Rise Vulnerability Assessment." February 2018. https://digitalcommons.humboldt.edu/hsuslri_local/33/.

Laird, Aldaron. 2018b. "Humboldt County Area Plan Sea Level Rise Vulnerability Assessment." January 2018. https://digitalcommons.humboldt.edu/hsuslri_local/17/.

Laird, Aldaron. 2018c. "King Salmon and Fields Landing: Communities at Risk Strategic Sea Level Rise Adaptation Planning" Presented at King Salmon and Fields Landing Workshop, August 7, Eureka, CA. Retrieved June 20, 202. https://humboldtgov.org/DocumentCenter/View/65366/King-Salmon-and-Fields-Landing-8-7-2018-Workshop-PowerPoint-PDF.

Laird, Aldaron. 2019. "Humboldt County Humboldt Bay Area Plan, Communities at-Risk Sea-Level Rise Vulnerability Assessment." Retrieved June 16, 2022. https://humboldtgov.org/DocumentCenter/View/81418/Humboldt-Bay-Area-Plan-Communities-at-Risk-SLR-Vulnerability-Assessment.

Laird, Aldaron, Brian Powell, and Jeff Anderson. 2013. "Humboldt Bay Shoreline Inventory, Mapping and Sea Level Rise Vulnerability Assessment." January 2013. https://digitalcommons.humboldt.edu/hsuslri_local/12/.

Laird, Aldaron, and Brian Powell. 2020. "Humboldt Bay Area Plan Consolidation of Key Sea Level Rise Information by Hydrologic Unit." Retrieved June 13, 2022. https://digitalcommons.humboldt.edu/hsuslri_geospatial/5/.

Lester, Charles. 2021. *Protecting Public Trust Shoreline Resources in the Face of Sea Level Rise.* Ocean and Coastal Policy Center, Marine Science Institute, University of California, Santa Barbara, California. https://documents.coastal.ca.gov/assets/slr/Lester%20Prot%20Public%20Trust%20Res%20Face%20of%20SLR.pdf.

Lindsey, Rebecca. 2022. "Climate change: Global Sea Level." *NOAA Climate*, April 19. Retrieved July 10, 2022. https://www.climate.gov/news-features/understanding-climate/climate-change-global-sea-level.

Lipe, Daniel J. 2013. "Diversifying Science: Recognizing Indigenous Knowledge Systems as Scientific Worldviews." PhD Dissertation, University of Hawai'i at Mānoa. Retrieved from ProQuest Dissertations & Thesis Database, 3572436.

Loveless, John P., and Brendan J. Meade. 2010. "Geodetic imaging of plate motions, slip rates, and partitioning of deformation in Japan." *Journal of Geophysical Research* 115(B2).

Marino, Elizabeth. 2015. *Fierce climate, sacred ground: An ethnography of climate change in Shishmaref, Alaska*. Fairbanks: University of Alaska Press.

Marino, Elizabeth. 2018a. "Adaptation privilege and voluntary buyouts: Perspectives on ethnocentrism in sea level rise relocation and retreat policies in the US." *Global Environmental Change* 49:10-13.

Marino, Elizabeth. 2018b. "Sea level rise and social justice: the social construction of climate change driven migrations." Pp. 181-193 in *Climate Change and Its Impacts: Risks and Inequalities,* edited by C. Murphy, P. Gardoni, and R. McKim. Switzerland: Springer International Publishing AG.

Massonnet, Didier, Thomas Holzer, and H. Vadon. 1997. "Land subsidence caused by the East Mesa geothermal field, California, observed using SAR interferometry." *Geophysical Research Letters* 24(8):901-904.

Matson, Laura, G. H. Crystal Ng, Michael Dockry, Madeline Nyblade, Hannah Jo King, Mark Bellcourt, Jeremy Bloomquist, Perry Bunting, Eric Chapman, Diana Dalbotten, Mae A. Davenport, Karen Diver, McKaylee Duquain, William Graveen, Katherine Hagsten, Kari Hedin, Susannah Howard, Thomas Howes, John Johnson, Shannon Kesner, Erik Kojola, Roger LaBine, Daniel J. Larkin, Melonee Montano, Seth Moore, Amy Myrbo, Michael Northbird, Meghan Porter, Rich Robinson, Cara M. Santelli, Riley Schmitter, Robert Shimek, Nancy Schuldt, Allison Smart, Donovan Strong, Joshua Torgeson, Darren Vogt, and Alexander Waheed. 2021. "Transforming Research and Relationships through Collaborative Tribal-University Partnerships on Manoomin (Wild Rice)." *Environmental Science & Policy* 115:108–15.

May, Christine L., Abigail T. Mohan, Olivia Hoang, Michael Mak, and Yanna Badet. 2020. "The Response of the Shallow Groundwater Layer and Contaminants to Sea Level Rise." Report by Silvestrum Climate Associates for the City of Alameda, California. doi:10.13140/RG.2.2.33390.69445.

Mazarrasa, I., N. Marbà, C. E. Lovelock, O. Serrano, P. S. Lavery, J. W. Fourqurean, H. Kennedy, M. A. Mateo, D. Krause-Jensen, A. D. L. Steven, and C. M. Duarte. 2015. "Seagrass meadows as a globally significant carbonate reservoir." *Biogeosciences* 12: 4993–5003.

McAdam, Jane. 2017. "Swimming against the tide: Why a climate change displacement treaty is not the answer." *International Journal of Refugee Law* 23(1):2–27.

McCaffrey, Robert, Anthony I. Qamar, Robert W. King, Ray Wells, Giorgi Khazaradze, Charles A. Williams, Colleen W. Stevens, Jesse J. Vollick, and Peter C. Zwick. 2007. "Fault Locking, Block Rotation and Crustal Deformation in the Pacific Northwest." *Geophysical Journal International* 169(3):1315–40

McGregor, Deborah. 2005. "Traditional Ecological Knowledge: An Anishnabe Woman's Perspective." *Atlantis: Critical Studies in Gender, Culture & Social Justice* 29(2):103-109.

Menzies, Charles R. 2006. *Traditional Ecological Knowledge and Natural Resource Management*. Lincoln: University of Nebraska Press.

Mohnot, Sona, Jordyn Bishop, and Alvaro Sanchez. 2019. *Making Equity Real in Climate Adaptation and Community Resilience Policies and Programs: A Guidebook*. Oakland, CA: The Greenlining Institute. August 2019. (https://greenlining.org/wp-content/uploads/2019/08/Making-Equity-Real-in-Climate-Adaption-and-Community-Resilience-Policies-and-Programs-A-Guidebook-1.pdf).

Nadasdy, Paul. 1999. "The Politics of Tek: Power and the 'Integration' of Knowledge." *Arctic Anthropology* 36(1/2):1–18.

Narayan, Siddharth, Michael W. Beck, Borja G. Reguero, Iñigo J. Losada, Bregje vanWesenbeeck, Nigel Pontee, James N. Sanchirico, Jane Carter Ingram, Glenn-Marie Lange, and Kelly A. Burks-Copes. 2016. "The Effectiveness, Costs and Coastal Protection Benefits of Natural and Nature-Based Defenses." *PLOS ONE* 11(5):e0154735. doi: 10.1371/journal.pone.0154735.

Newkirk, Sarah, Sam Veloz, Maya Hayden, Bob Battalio, Tiffany Cheng, Jenna Judge, Walter Heady, Kelly Leo, and Mary Small. 2018. *Toward Natural Shoreline Infrastructure to Manage Coastal Change in California*. CCCA4-CNRA-2018-011. California Natural Resources Agency.

Newsom, Gavin. 2020. *Executive Order N-82-20*. State of California. Retrieved June 30, 2022. https://www.gov.ca.gov/wp-content/uploads/2020/10/10.07.2020-EO-N-82-20-.pdf.

Newsom, Office of Governor. 2022. Governor Newsom Signs Budget Putting Money Back in Californians' Pockets and Investing in State's Future. Published: June 30, 2022. Retrieved January 2023. https://www.gov.ca.gov/2022/06/30/governor-newsom-signs-budget-putting-money-back-in-californians-pockets-and-investing-in-states-future/

Nicholls, Robert J. 2011. "Planning for the Impacts of Sea Level Rise." *Oceanography* 24(2):144–157.

Nicholls, Robert J., and Anny Cazenave. 2010. "Sea-Level Rise and Its Impact on Coastal Zones." *Science* 328(5985):1517–20.

NHE (Northern Hydrology and Engineering) and Jeffrey K. Anderson. 2014. *Estimates of Local or Relative Sea Level Rise for Humboldt Bay Region*. Prepared for the California State Coastal Conservancy and Coastal Ecosystems Institute of Northern California.

OPC (California Ocean Protection Council). 2018. *State of California Sea Level Rise Guidance: 2018 Update.* California Natural Resources Agency and Ocean Protection Council. https://www.opc.ca.gov/webmaster/ftp/pdf/agenda_items/20180314/Item3_Exhibit-A_OPC_SLR_Guidance-rd3.pdf)

OPC (California Ocean Protection Council). 2020. *Strategic Plan to Protect California's Coast and Ocean 2020–2025*. Retrieved August 15, 2022. https://www.opc.ca.gov/webmaster/ftp/pdf/agenda_items/20200226/OPC-2020-2025-Strategic-Plan-FINAL-20200228.pdf

OPC (California Ocean Protection Council). 2022. *Ocean Protection Council Equity Plan.* October 2022. https://www.opc.ca.gov/webmaster/_media_library/2022/11/OPC-Equity-Plan-508.pdf.

OPC (California Ocean Protection Council). 2023. *Tribal Engagement Strategy*. January 2023. https://www.opc.ca.gov/webmaster/ftp/pdf/agenda_items/20230124/OPC-Tribal-Engagement-Strategy-508.pdf.

Oppenheimer, Michael, Bruce C. Glavovic, Jochen Hinkel, Roderik Van de Wal, Alexandre K. Magnan, Amro Abd-Elgawad, Rongshuo Cai, Miguel Cifuentes-Jara, Robert M. DeConto, Tuhin Ghosh, John Hay, Federico Isla, Ben Marzeion, Benoit Meyssignac, and Zita Sebesvari. 2022. "Sea Level Rise and Implications for Low-Lying Islands, Coasts and Communities." Pp. 321–445 in *IPCC Special Report on the Ocean and Cryosphere in a Changing Climate*, edited by A. Abe-Ouchi, K. Gupta, and J. Pereira. Cambridge, UK and New York, NY: Cambridge University Press.

Orth-Gordinier, Kristen. 2022. "Social Science Research to Help Advance Regional Coordination and Collaboration of Sea Level Rise Planning and Adaptation on Humboldt Bay." Master's Thesis, Natural Resources: Environmental Science and Management, Cal Poly Humboldt.

Patton, Jason R., Todd B. Williams, Jeff Anderson, Thom H. Leroy, Kyle Weiss, Reed Burgette, Ed Southwick, Whelan Gilkerson, Eric Nelson, Jay Stallman, Susan Schlosser, Mark Hemphill-Haley, Diane Sutherland, Ray Weldon II, and student volunteers. 2017. "Tectonic land level changes and their contribution to sea-level rise, Humboldt Bay region, Northern California." Final Report for US Fish and Wildlife Service Agreement #81331BJ244. http://www.hbv.cascadiageo.org/HumBayVert/reports/USFWS/20170407/final_report_HBV_usfws_2017.pdf.

Patton, Jason R., Todd B. Williams, Jeffrey K. Anderson, Mark Hemphill-Haley, Reed J. Burgette, Ray Weldon II, Robert C. McPherson, and Thomas H. Leroy. 2023. "20th to 21st Century Relative Sea and Land Level Changes in Northern California: Tectonic Land Level Changes and Their Contribution to Sea-Level Rise, Humboldt Bay Region, Northern California." *TeKtoniKa* 1(1). doi:10.55575/tektonika2023.1.1.6.

Pearce, Fred. 2022. "Why Are Nature-Based Solutions on Climate Being Overlooked?" *Yale E360,* April 18. Retrieved July 22, 2022. https://e360.yale.edu/features/why-are-nature-based-solutions-on-climate-being-overlooked.

Peek, Lori, Jennifer Tobin, Rachel M. Adams, Haorui Wu, and Mason Clay Mathews. 2020. "A Framework for Convergence Research in the Hazards and Disaster Field: The Natural Hazards Engineering Research Infrastructure CONVERGE Facility." *Frontiers in Built Environment* 6(110). doi:10.3389/fbuil.2020.00110.

Pelling, Mark. 2010. *Adaptation to Climate Change: From Resilience to Transformation*. London, Routledge.

Peltier, W. R. 1976. "Glacial-Isostatic Adjustment – II. The Inverse Problem." *Geophysical Journal International* 46(3):669–705.

Perumal, N. (2018). "'The place where I live is where I belong': community perspectives on climate change and climate-related migration in the Pacific island nation of Vanuatu." *Island Studies Journal*, 13(1):45–64.

Pierre-Louis, Kendra. 2021. "How Rising Groundwater Caused by Climate Change Could Devastate Coastal Communities." *MIT Technology Review*. December 13. Retrieved December 13, 2022. https://www.technologyreview.com/2021/12/13/1041309/climate-change-rising-groundwater-flooding/.

Plumer, Brad. 2020. "Rising Seas Could Menace Millions Beyond Shorelines, Study Finds." *The New York Times*, July 30. Retrieved July 10, 2022. https://www.nytimes.com/2020/07/30/climate/sea-level-inland-floods.html.

Polk, Merritt. 2014. "Achieving the Promise of Transdisciplinarity: A Critical Exploration of the Relationship between Transdisciplinary Research and Societal Problem Solving." *Sustainability Science* 9(4):439-451.

Polk, Merritt. 2015. "Transdisciplinary Co-Production: Designing and Testing a Transdisciplinary Research Framework for Societal Problem Solving." *Futures* 65:110–22.

Prosser, Diann J., Thomas E. Jordan, Jessica L. Nagel, Rochelle D. Seitz, Donald E. Weller, and Dennis F. Whigham. 2018. "Impacts of Coastal Land Use and Shoreline Armoring on Estuarine Ecosystems: An Introduction to a Special Issue." *Estuaries and Coasts* 41(Suppl 1):S2–18.

Ramos, Seafha C. 2022. "Understanding Yurok Traditional Ecological Knowledge and Wildlife Management." *The Journal of Wildlife Management* 86(1):e22140. doi:10.1002/jwmg.22140.

Reed, Kaitlin. 2022. "Creating a PolyTEK," *Humboldt. The Magazine of Cal Poly Humboldt,* Fall 2022, p. 33.

Reibold, Kerstin. 2022. "Settler Colonialism, Decolonization, and Climate Change." *Journal of Applied Philosophy*.

Reid, Walter V., Fikret Berkes, Thomas Wilbanks, and Doris Capistrano. 2006. *Bridging Scales and Knowledge Systems: Concepts and Applications in Ecosystem Assessment.* Island Press. doi:10.1111/japp.12573.

Resilience Alliance. 2022. "About." Retrieved February 10, 2023. https://www.resalliance.org/about.

Risling Baldy, Cutcha, Kayla Begay, and Kaitlin Reed. 2023. "Polytech to PolyTEK: Traditional Ecological Knowledge, Indigenous Science, and the Future Forward Polytechnic University." In M. Virnoche, J. Eichstedt, E. Kelly, & K. Morgan (Eds.), *Humboldt Journal of Social Relations* 1(45):34-51. The Press at Cal Poly Humboldt.

Rittel, Horst W. J., and Melvin M. Webber. 1973. "Dilemmas in a General Theory of Planning." *Policy Sciences* 4(2):155–69.

Rohde, Jerry. 2020. *Humboldt Bay Shoreline, North Eureka to South Arcata: A History of Cultural Influences*. Arcata, CA: Humboldt State University Press. https://digitalcommons.humboldt.edu/cgi/viewcontent.cgi?article=1022&context=hsuslri_local.

Rotzoll, Kolja, and Charles H. Fletcher. 2013. "Assessment of Groundwater Inundation as a Consequence of Sea-Level Rise." *Nature Climate Change* 3(5):477–81.

Ryan, John. 2022. "Swinomish Tribe Builds Modern Clam Garden, Reviving Practice." *The Seattle Times*, September 4. Retrieved June 2, 2022. https://www.seattletimes.com/seattle-news/swinomish-tribe-builds-modern-clam-garden-reviving-practice/.

Shaughnessy, Frank J., Whelan Gilkerson, Jeffrey M. Black, David H. Ward, and Mark Petrie. 2012. "Predicted Eelgrass Response to Sea Level Rise and Its Availability to Foraging Black Brant in Pacific Coast Estuaries." *Ecological Applications* 22(6):1743–61.

Shaughnessy, F. J., Mulligan, T., Kramer, S., Kullmann, S., and J. Largier. 2017. "Baseline characterization of biodiversity and target species in estuaries along the North Coast of California." Final Report for California's North Coast Marine Protected Area Baseline Monitoring Program Project #R/MPA-40A. doi:10.13140/RG.2.2.30944.94724.

Short, Frederick T., and Hilary A. Neckles. 1999. "The Effects of Global Climate Change on Seagrasses." *Aquatic Botany* 63(3–4):169–96.

Siders, A. R., and Idowu Ajibade. 2021. "Introduction: Managed Retreat and Environmental Justice in a Changing Climate." *Journal of Environmental Studies and Sciences* 11(3):287–93.

SLRI (Cal Poly Humboldt Sea Level Rise Institute). 2023. "Maps." Retrieved March 10, 2023. https://humboldtslri.org/.

Smith, Linda Tuhiwai. 2021. *Decolonizing Methodologies: Research and Indigenous Peoples*. 3rd ed. London: Zed Books.

Swanson, Kayleigh. 2021. "Equity in Urban Climate Change Adaptation Planning: A Review of Research." *Urban Planning* 6(S2):287–97.

Tetra Tech. 2019. "Humboldt County Operational Area Hazard Mitigation Plan 2019." Vol. 1, *Area-Wide Elements*. Public Review Draft. Retrieved July 7, 2022. https://digitalcommons.humboldt.edu/cgi/viewcontent.cgi?article=1032&context=hsuslri_local.

The Land Institute. 2022. "Civic Science." Retrieved January 25, 2023. https://landinstitute.org/our-work/ecosphere-studies/civic-science/.

Vanderheiden, Isabella. 2022. "Coastal Commission Reviews Sea Level Mitigation Measures for Arcata-Eureka Highway 101 Corridor." *Lost Coast Outpost*, July 15. Retrieved July 10, 2022. https://lostcoastoutpost.com/2022/jul/15/coastal-commission-reviews-sea-level-mitigation-me/.

Vaughan, Mehana Blaich. 2018. *Kaiāulu: Gathering Tides*. Corvallis: Oregon State University Press.

Wang, Kelin, Jiangheng He, Herb Dragert, and Thomas S. James. 2001. "Three-Dimensional Viscoelastic Interseismic Deformation Model for the Cascadia Subduction Zone." *Earth, Planets, and Space* 53(4):295–306.

Weber, Edward P., D. Lach, and Brent Steel. 2017. *New Strategies for Wicked Problems: Science and Solutions in the Twenty-First Century*. Corvallis, OR: Oregon State University Press.

WLI (Water Leaders Institute). 2022. "Water Leaders Institute." New Orleans, LA. Retrieved July 2022. https://www.waterleadersinstitute.com.

Willis, Robert. 2014. "Conceptual Groundwater Model of Sea Level Rise in the Humboldt Bay Eureka-Arcata Coastal Plain." Final Report for California State Coastal Conservancy and Coastal Ecosystems Institute of Northern California. December 2014. https://humboldtbay.org/sites/humboldtbay2.org/files/Final_HBSLR_ConceptualGroundwaterModel_141209.pdf.

Wiyot Tribe. 2022. "History" Retrieved July 22, 2022. https://www.wiyot.us/148/Cultural.

Wyllie-Echeverria, Sandy, Karl Gunnarsson, Miguel A. Mateo, Joseph A. Borg, Pinsach Renom, John Kuo, Anja Schanz, Frida Hellblom, Emma Louise Jackson, Gerard Pergent, Christine Pergent-Martini, Monique Johnson, J. Sanchez-Lizaso, Charles Boudouresque, and Keiko Aioi 2002. "Protecting the Seagrass Biome: Report from the Traditional Seagrass Knowledge Working Group." *Bulletin of Marine Science* 71(3):1415–17.

Appendix

To view the appendix for this article by scanning the QR code below or following this link: https://digitalcommons.humboldt.edu/cgi/viewcontent.cgi?article=1005&context=oer.

HJSR
50TH ANNIVERSARY EDITION

Smoke, Air, Fire, Energy (SAFE) in Rural California: Critical Reflections on an Interdisciplinary Research Collaboration

Deepti Chatti *<dc317@humboldt.edu> (Cal Poly Humboldt),* ***Carisse Geronimo*** *(Schatz Energy Research Center),* ***Cassidy Barrientos*** *(Schatz Energy Research Center),* ***Jana Ganion*** *(Blue Lake Rancheria),* ***Malcolm Moncheur*** *(Cal Poly Humboldt),* ***Peter Alstone*** *(Cal Poly Humboldt),* ***Shawn "Shay" Bourque*** *(Karuk Department of Natural Resources),* ***Tanya Garcia*** *(Schatz Energy Research Center),* ***Tesfayohanes Yacob*** *(Cal Poly Humboldt)*

Abstract

This article provides a synthesis of the interconnected problems of tenuous energy access, wildfires, and exposures to high air pollution in Indigenous communities in rural California through the lens of ongoing collaborative research being carried out by researchers at Cal Poly Humboldt, Schatz Energy Research Center, Karuk Department of Natural Resources, and the Blue Lake Rancheria Tribe. The collaboration is funded by the Strategic Growth Council of the state of California, and we hope is the beginning of a longer term relationship between all partners. We are an interdisciplinary team of researchers drawing on energy engineering, air pollution science, and qualitative social sciences to better understand the intersecting challenges of expanding clean energy access, and building climate resilience in Tribal communities in rural California in the context of the multiple challenges of climate change, increasing risk of dangerous wildfires, and high exposures to air pollution. Individuals and communities need to make decisions about energy and air quality infrastructure with implications for public health, climate change, energy resilience, and Tribal sovereignty. This article will reflect on the joys, challenges, ethical questions, and epistemological constraints involved with academic researchers working on interdisciplinary research projects across disciplines, and in partnership with Tribal nations. Grounded in the reflections and experience of an ongoing project, this article sheds light on the challenges and unique opportunities of conducting collaborative interdisciplinary research in close engagement with communities, and also reflects on the structural constraints posed within current institutional structures.

Keywords: wildfires, smoke, energy, climate change, Tribes, California

Wildfires, smoke, and energy access are intricately linked in Northern California. While California has always been a fire-adapted place (Pyne 2016), recent years have seen increasing incidences of catastrophic wildfires due to a constellation of reasons including climate change. In the last five years in California, 10 million acres have burned in wildfires and 151 human lives have been lost (CALFIRE 2022). These fires are linked not only with dangerous conditions for people, wildlife, and ecosystems, leading to tragic loss of human and non-human lives, homes, and habitats, but also cause extremely high air pollution levels in the region. Simultaneously, the electric utility serving much of the region, Pacific Gas and Electric (PG&E), manages fire risk by powering down the electric grid during conditions deemed unsafe and prone to spark wildfires. This practice means that residents in the region do not have electricity to run air filters when these technologies are most needed. As it is, rural and Tribal communities face precarious access to the grid (Sandoval 2018), and the quality of their energy services is made worse with the expanding fire season and the subsequent blackouts.[1] Wildfires become catastrophic with fuel accumulation on the landscape due to settler-colonial policies that criminalize Indigenous land management practices that use fire as a tool to tend to the landscape (Tripp 2020, Norgaard 2014, Norgaard 2019). And finally, this dynamic set of issues unfolds in a context where the state of California has committed to meeting its full electricity requirements from renewable and zero-carbon energy sources by 2045 (as laid out in Senate Bill 100), requiring an expansion and strengthening of its electricity infrastructure.

To begin addressing the nexus of issues surrounding wildfires, smoke, and energy access in Northern California, a diverse group of academics and practitioners came together to conceptualize a research collaboration set in Humboldt County. The Smoke, Air, Fire, Energy (SAFE) collaboration is a result of that engaged process, and is an ongoing interdisciplinary, community-engaged, research partnership co-led by faculty at Cal Poly Humboldt, the Schatz Energy Research Center, the Karuk Tribe, and Blue Lake Rancheria Tribe, with an overall goal of increasing energy and climate resilience in Tribal and rural California communities. The focus of the collaboration is on analyzing the interconnections between three entangled issues: wildfires of increasing intensity and frequency in Northern California, concomitant worsening air pollution levels, and precarious energy access faced by rural and Tribal communities. The collaboration goals include understanding the interconnections of these issues; foregrounding the political sovereignty, knowledge, and expertise of Tribes on fires; using new technologies to understand the air pollution impacts of fires; and making progress towards strengthening energy infrastructures in historically underserved Native and rural communities in the context of climate change.

As a community-engaged research collaboration, this program was conceptualized from its very beginning in conversations between the Karuk Tribe, Blue Lake Rancheria, and faculty from Cal Poly Humboldt. As an interdisciplinary research collaboration, this intellectual effort draws on varied disciplinary perspectives from academia including faculty in engineering and faculty in the critical social sciences. The epistemic and political orientation of the collaboration includes honoring Tribal sovereignty and expertise, and integrating Indigenous knowledge actively held and practiced by the Karuk Tribe, institutional and technical expertise held by the Blue Lake Rancheria with regards to energy resilience, and technoscientific approaches to air quality monitoring and clean energy, with an overall goal of enhancing the energy and air quality infrastructure in the region. Following feminist science and technology studies (STS) scholars who have pushed scientists and other producers of knowledge to "re-make" science, center questions of power, and interrogate what it means to conduct empirical research (see for example Subramaniam and Willey 2017; Liboiron 2021; Goldman, Nadasdy and Turner 2011), we have grounded our collaboration's questions around air quality and energy infrastructures in the

1. From 2014-2018, customers in Orleans experienced four times more outage minutes than the average PG&E customer (PG&E 2018).

knowledge of the historic and contemporary struggles of Indigenous peoples to practice their expertise and knowledge amidst the onslaught of colonial policies that actively perpetrate violence to erase Indigenous peoples and their knowledge from the land. From this starting point, practicing science as "neutral," or as not having a point of view (Haraway 1988), further perpetuates this erasure by obscuring its violent history. Our collaboration makes visible our political stakes in this process by actively supporting the needs of the Karuk Tribe and Blue Lake Rancheria Tribe.

Conceptualizing a cohesive research collaboration on these vast topics in a way that could be neatly articulated to a funding agency was no small undertaking. We were keen, from the outset, that this research collaboration be driven by community needs. We co-wrote and edited the proposal through an iterative process with all collaborators. Furthermore, the extremely different disciplinary genres of knowledge production that our diverse group engages with (and is accountable to) in our academic lives meant that we had to create new ways of producing knowledge together. In simple terms, we wanted this to be neither an engineering, air pollution science, or critical social science project in its entirety, nor did we want it to be three separate pieces of research under one program umbrella. We wanted to create something that drew on each of these ways of understanding the world and to have each component inform the others. To make this possible, we wanted to go beyond working in our distinct areas separately, and do more than speaking to each other now and then about how our individual research areas were coming along. Critical social scientists reading this article will appreciate the challenge of conducting research in a way that could inform engineering designs and air pollution sensor networks, and engineers and scientists reading this article will appreciate the challenge of including a consideration of political sovereignty and multiple epistemic frameworks while trying to size a microgrid or monitor particulate matter concentrations in the air. Academics of all disciplines will appreciate the challenges for junior faculty to meet tenure expectations while simultaneously building relations of trust with community partners whose priorities and needs may be substantively different than those of our colleagues within academia. We do not claim to have neatly articulated answers on how to do this; we do not think that we have solved interdisciplinarity nor resolved the tensions between the expectations of academic knowledge production and our collaborators' pragmatic needs outside it. But we offer our reflections on the challenges and opportunities of carrying out such work in collaboration with each other on pressing research topics of deep concern to the communities in which Cal Poly Humboldt is situated. It is our hope that our critical and candid reflections might inspire others to embark on similarly ambitious intellectual pursuits even when they are messy. Perhaps informed by our attempts and what we have learned, we can collectively and more effectively address contemporary environmental and societal challenges.

Research Context

This ongoing collaboration is located in Humboldt County in rural Northern California, about 300 miles north of the San Francisco Bay Area. Cal Poly Humboldt is located in Arcata and is part of the California State University. Cal Poly Humboldt holds an institutional vision to "be the premier center for the interdisciplinary study of the environment, climate crisis and resilience to climate change, and the conservation of ecological systems and natural resources" and to "partner with Indigenous communities to address the legacy of colonialism, and create space nurturing of traditional ecological knowledge, pedagogies, and curricula responsive to their identified needs" (Cal Poly Humboldt Vision 2022).

The Blue Lake Rancheria is a federally recognized Tribe whose vision is "to secure a better future for its people; protect its sovereignty and heritage; learn from the past; and build a resilient, healthy economy and environment, with benefits for the Tribe, the region, and the planet." The Karuk Tribe is located approximately 100 miles inland from the coast in ancestral territories in the mid-Klamath region. Prominent towns in Karuk territories include Orleans and Happy Camp. This region of the world experiences some of the worst air quality in the continental United States due to wildfires (Ford et al, 2018), and also experiences tenuous electricity service from electric utilities.

Details of the Interdisciplinary, Collaborative Research Partnership

Our interdisciplinary, collaborative research partnership developed over a number of years. Between 2014-2019, the Blue Lake Rancheria Tribe collaborated extensively with the Schatz Energy Research Center on numerous projects related to energy. These projects involved conceptualizing, designing, and building clean energy infrastructure systems, and established trust between the Tribe, the Schatz Energy Research Center, and faculty at Cal Poly Humboldt. In 2019, with increasing concerns around wildfire smoke in the region, conversations were begun regarding partnering on research on air quality as well, with a desire to explore the feasibility of using air filters to protect residents in the region during wildfire events. Since the air quality inland in Karuk ancestral territories is far worse than the air pollution levels experienced on the coast around Blue Lake Rancheria and Cal Poly Humboldt, a conversation was also started with the Karuk Department of Natural Resources to explore the possibility of expanding the collaboration to include the Karuk Tribe as a research partner.

Undergraduate students from Cal Poly Humboldt studying engineering were also included in this early stage. As part of their senior capstone project, students worked on air quality infrastructure designs useful for the Karuk and Blue Lake Rancheria communities. These included designing air filtration systems for community buildings and residences, developing plans for a sensor network, and considering management plans for community air quality. The student teams identified a number of themes that continue to inform the SAFE work in their designs: the need for much larger air filtration flow rates than are typical from commercially available systems to manage heavy smoke inundation, the value of low-cost sensors for situational awareness, and the difficulty of maintaining air quality infrastructure.

As these conversations and student projects were underway, the first "public safety power shutoff" events affected the region. PG&E de-energized the grid as a way to reduce the risk of the electricity infrastructure sparking wildfires. Faculty in the critical social sciences were keen to study the inequalities experienced by various communities in accessing the energy grid, and the inequalities in air pollution exposures as a result of historic and contemporary marginalization. Researchers across disciplines initiated conversations about studying multiple facets of these complicated and dynamic issues that were rapidly unfolding. These interdisciplinary conversations further emphasized the connections between wildfires, Tribal sovereignty, energy and electricity infrastructure, community perceptions of and responses to air pollution, and air quality infrastructures for monitoring and mitigation of air pollution. The SAFE collaboration grew out of these conversations.

A grant opportunity from the Strategic Growth Council of the Governor's Office of the State of California presented itself. The team engaged in an iterative writing process including academic and Tribal partners, and drawing on the expertise of those involved. The process included engaging in conversations among all the partners, face-to-face and phone meetings of small groups at Cal Poly Humboldt, writing together as a large team in Orleans at the Karuk Department of Natural Resources (the document was projected on a large screen for editing), and sharing drafts over email and Google Drive for commenting, editing, revising, and rewriting. The final grant application was submitted early in 2019. In Spring 2020 we received funding for the collaboration.

Since we are co-producing knowledge, and not placing any one form of knowledge production or any one discipline of academia in a position of power over another form of knowledge production or academic discipline, the research partnership is a flat structure with five Principal Investigators (PIs) from across locales and expertise areas at the university and in the Tribes (see figures 1 and 2). The PIs are responsible for leading the intellectual work of the project, and additional research staff at the Schatz Energy Research Center and undergraduate and graduate students from Cal Poly Humboldt have been actively involved in carrying out research activities for varying lengths of time.

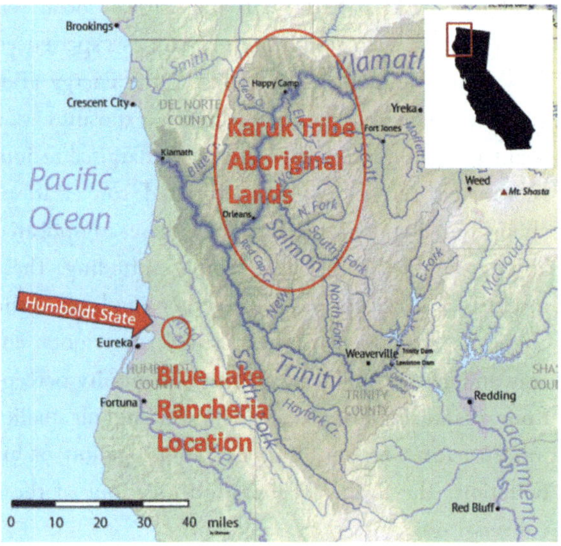

Figure 1: Location of the collaborative partners in Humboldt County; inset shows location of the partners in far Northern California

Research Objectives and Questions

To achieve the collective goals of the collaboration, the research team divided the intellectual work into three distinct overarching objectives within which research questions were formulated to be answered. The main objectives are:

***Objective A*: Accelerating climate-smart energy and air quality infrastructures**

We are developing engineering design tools, management plans, and financial strategies that accelerate deployment of energy and air quality infrastructure (that is identified as needed by the collaborative process) at three scales: households, critical facilities, and isolated community clusters of 10-50 households and businesses.

***Objective B*: Understanding social dimensions of change**

We are working with community members and leaders to advance understanding of the social dimensions of climate-smart and fire-smart infrastructures and practices.

***Objective C*: Advancing sustainable university-community research partnerships**

We are identifying the institutional needs and opportunities for universities to play a supportive, long-term role in connecting indigenous communities in their region with resources for environmentally just community development and research.

These overarching objectives were divided into research questions that would need to be answered to achieve the objective. A subset of research questions was on the more technical end of the spectrum, with their outcomes derived from quantitative calculations of energy systems and air quality data. Another subset of research question was more critical in nature and their outcomes were derived from qualitative data collected using semi-structured interviews, focus group discussions, and participant observations. Yet another set of research question were focused on the process of the collaboration itself, and self-reflexively analyzed the institutional challenges faced by the collaborators in working together on these topics.

Research Questions

1. How can we design clean energy microgrids that are resilient to wildfire-induced risk and serve critical needs at three scales: households, critical facilities, and isolated community clusters of 10-50 households and businesses?
2. How can we use low-cost air quality sensing networks and indoor air filtration systems in supporting healthy rural communities and healthy forests across the landscape?
3. How can people's understanding of smoke, air, fire, and energy help inform our approaches to managing infrastructure systems in support of climate goals and sustainable landscapes?
4. How can research partnerships between universities and the communities they are a part of transcend one-off reports and assessments, resulting instead in meaningful and sustained collaboration?

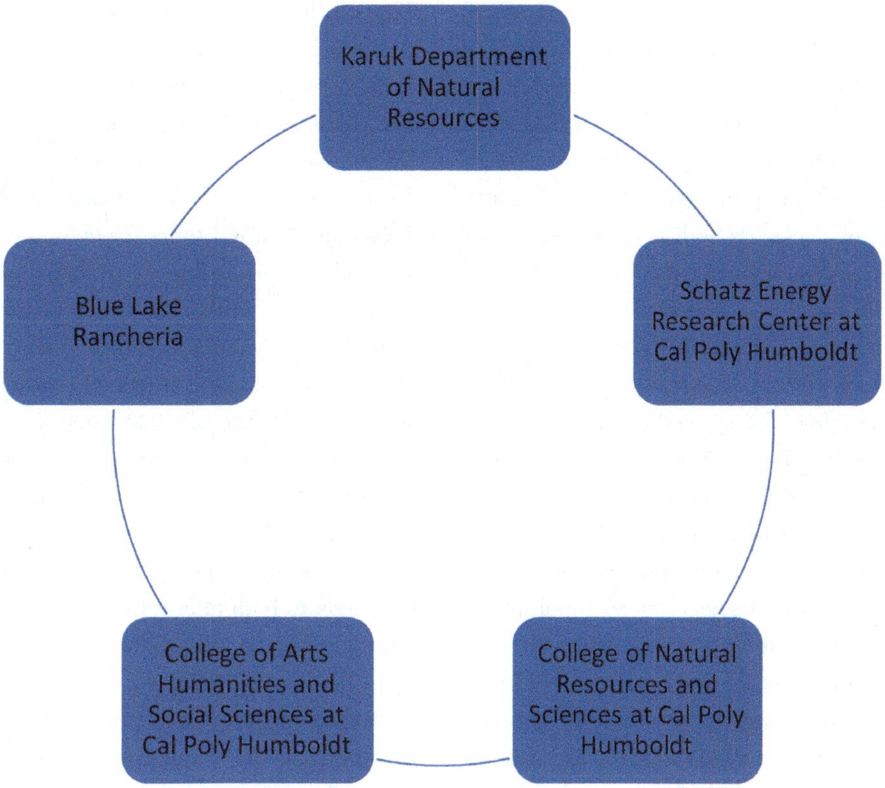

Figure 2: Primary institutional affiliations of the five Principal Investigators (PIs) of the collaboration

As we mentioned in an earlier section, these questions emerged and were developed through a series of iterative conversations. The way we approached working on this project was built on ensuring that lines of communication stayed open and active. We set up weekly virtual discussions where our collaboration team parsed out, reframed, and reconsidered many of these questions in the early parts of our work. A s a group, we decided to prioritize opportunities to impact policy and direct material resources to meet community needs. This decision meant that we prioritize writing grant proposals over research papers, and seek opportunities for strategic discussions with policymakers. These happen throughout the collaboration process rather than hoping someone will eventually read our final report. Developing follow-on infrastructure grants and setting up meetings with officials precipitated and motivated engagement with a range of people in the Tribal governments and community in their governance, personal, and cultural roles. It helped to make the research collaboration work towards tangible and specific infrastructure goals, which also let us reveal the cracks in the systems that would deliver that infrastructure.

Initial Projects of the Collaborative Team

While this work is an ongoing collaboration and we are actively collecting data on multiple research questions and engaging in different projects, we are able to share the following preliminary insights and outcomes from our activities. We also note that some of these activities are further along than others, and thus, some of our following sections are more developed than others. Where the research task is completed, we have so indicated in the subsection.

Improving Understanding of Smoke During Fire Events

The collaboration team relies on multiple ways of collecting information about smoke including semi-structured interviews with community residents, focus group discussions, and air quality monitoring and analysis using sensors. Many community members living in the Karuk ancestral territories have noted that the smoke during wildfire season has become increasingly worse in recent years in quality, intensity, and duration. The worsening of smoke has both temporal and spatial dimensions. Residents described that even 20-30 years ago, they would experience "bad air days" for only one week in the summer, whereas more recent summers have had such days for over two months. Spatially, residents described being able to "escape to the coast" in decades past, but now feel a sense of entrapment in the bad air.

Mountainous terrain, differences in elevation, wind directions, and proximity to fires all lead to highly variable levels of smoke across the landscape changing throughout the day. The installation of multiple air sensors at key locations in the region enhance our fine-grained understanding of the way smoke moves in the area during fire events, and are now used by residents of the Orleans-Somes Bar region to make logistical decisions regarding keeping windows open or closed, running air filters, wearing masks, or evacuating. As part of the SAFE collaboration, 11 "Purple Air" sensors have been installed so far across the region and are providing continuous data that is tracked by researchers (see figures 3 and 4). These sensors monitor levels of small particulate matter called PM 2.5, which is a key constituent of wildfire smoke and a pollutant of concern (Wagner and Chen 2019). The data from these sensors is available for anyone with access to the internet. The expanded sensor network is already used by staff at the Karuk Department of Natural Resources and some residents of the region to understand the smoke status.

While understanding smoke is extremely important during wildfire events, we are also keen to better understand the patterns of smoke from prescribed fires and cultural burns. In this article we are using the term "prescribed fires" to refer to intentional fires that are used to manage forests and grasslands for reducing the accumulation of flammable fuels in the landscape near communities. We are using 'cultural burns' to refer to intentional fires used for improving the health of specific types of plants determined as culturally significant by the Karuk Tribe. The purpose of smoke monitoring around prescribed and cultural fires is to support the Karuk Tribe's efforts to "*put good fire back on the land*", a desire expressed nearly universally by everyone interviewed for the collaboration. There is widespread support in the region for more actively managing the landscape in Karuk ancestral territories for two purposes: increasing the health of plants that are culturally important to the Karuk Tribe, as well as preventing catastrophic wildfires in the region by not letting fuels build up to high levels in the landscape. Both anecdotal experience and emerging research (Jaffe et al. 2020, Long et al. 2018, Prunicki et al. 2019) suggest that smoke levels during intentional fires are less intense and of shorter duration than wildfires. Karuk Tribe practitioners indicated that ongoing monitoring of air quality could help in the short-and long-term to both inform their work and also make the case to the community and regulators.

In October 2021, we measured smoke levels in close proximity to prescribed and cultural fires, building on work that the Karuk Tribe has done in the past to document the relatively low impact of prescribed burns compared to wildfires (Tripp 2017). The photos in Figure 5 show a prescribed burn from the 2021 Klamath TREX, which included operations to protect U.S.F.S. controlled areas in the "Patterson" burn unit (Western Klamath Restoration Partnership 2022). The data from the temporary smoke sensor installed just to the south of the fire operations, displayed in Figure 6, shows that community members living close to the fire operation experienced somewhat elevated, but not acutely hazardous, levels of smoke.

A cultural fire near Orleans on October 15, 2021 was focused on willows that were overgrown. Willows are significant as basket materials for the Karuk Tribe. Experts from the Karuk Tribe collaborated with the Prescribed Fire Training Exchanges (TREX) program for this burn event, and temporary smoke sensors were installed for monitoring, both in the main part of Orleans and close to housing just across the river (see figure 7). Data from the sensors, displayed in Figure 8, indicate that there was a short period of unhealthy to very unhealthy smoke levels in one part of Orleans, but only moderate levels in another neighborhood. This empha-

Figure 3: Smoky skies in Orleans, CA and a map of widespread wildfire smoke on August 31, 2021. Project areas circled in white.

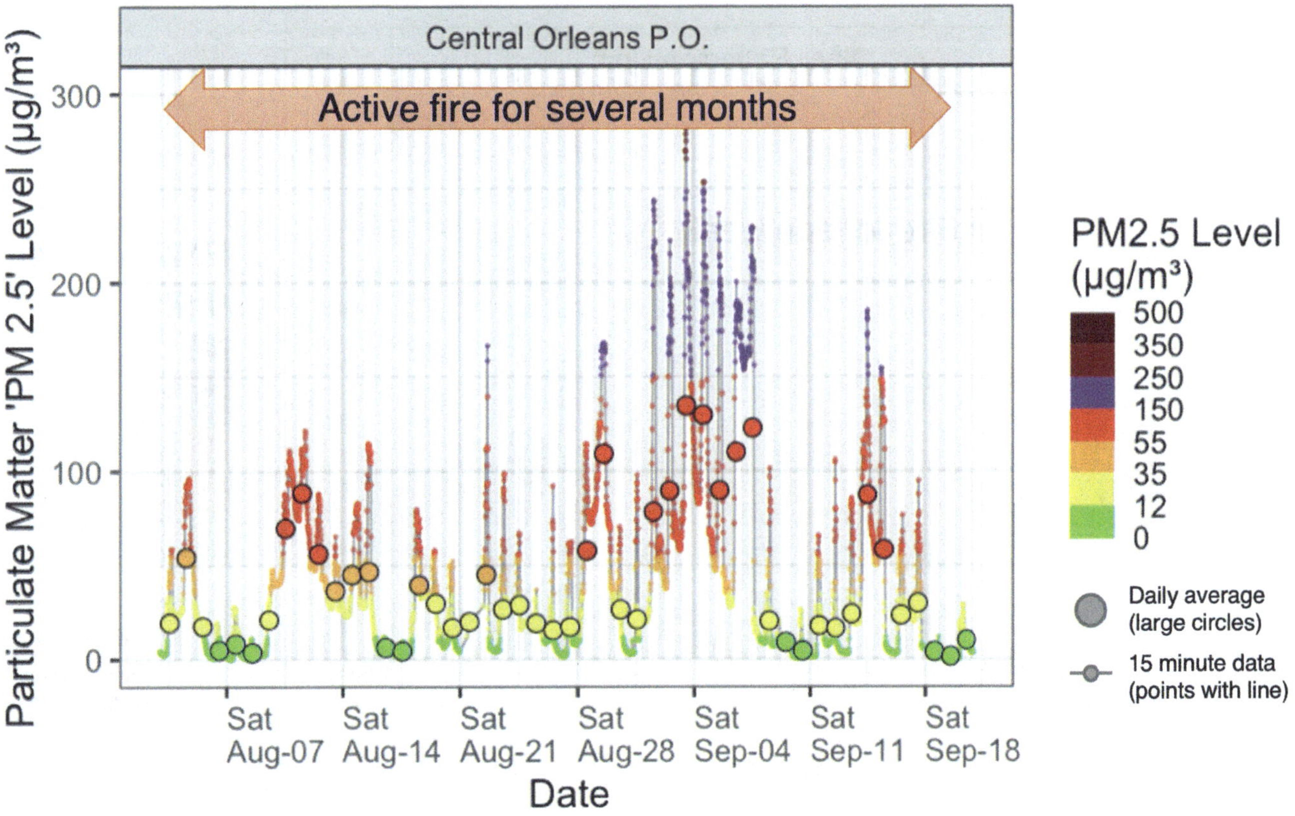

Figure 4: PM 2.5 levels in central Orleans, CA

Figure 5: (Left) Prescribed fire in the "Patterson" burn unit in Six Rivers National Forest in October 2021. (Right) A temporary air quality monitor powered by solar energy that was installed near the burn.

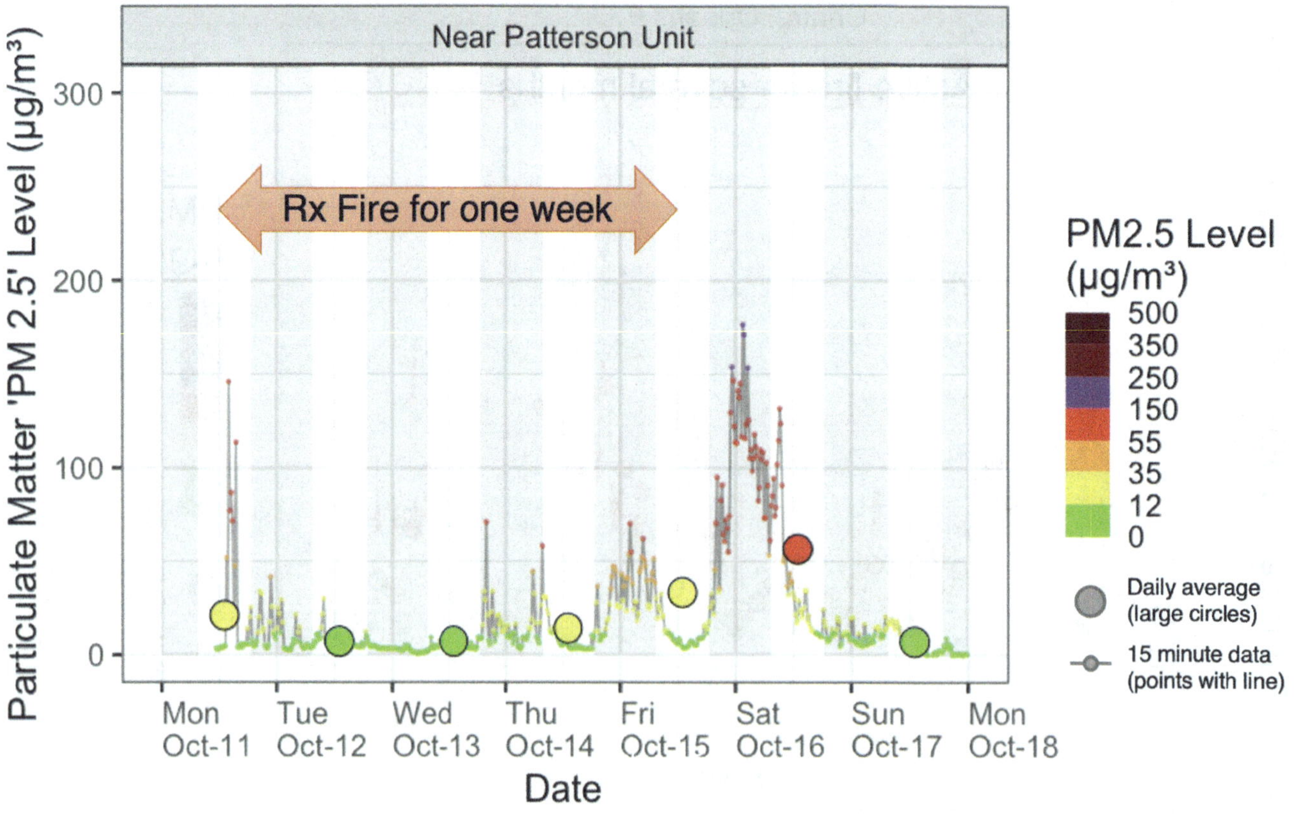

Figure 6: Smoke (PM 2.5) levels as measured by a temporary monitor near a prescribed fire in the Patterson Unit of the Somes Bar Integrated Fire Management Project in October 2021.

Figure 7: Figure shows fire practitioner igniting brush near the willows and a temporary smoke sensor at a cultural burn in October 2021.

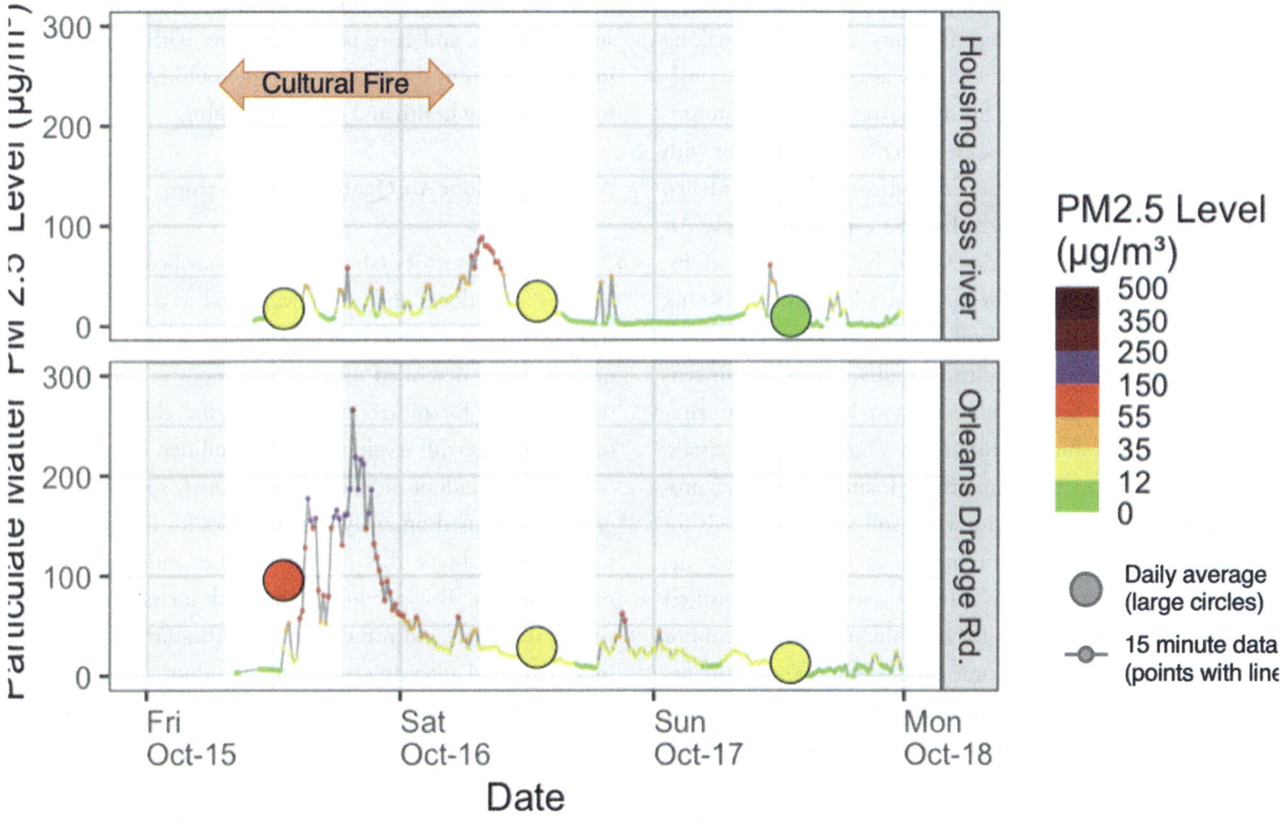

Figure 8: Smoke levels at two locations near the cultural burn in October 2021.

sizes the importance of localized monitoring to understand the smoke impacts from cultural and other intentional fires.

In addition to the variation in smoke levels associated with intentional fires and wildfires, it is noteworthy that all smoke is not considered the same qualitatively. Based on visual, olfactory, taste, and other sensory cues, community members differentiate between different kinds of smoke. For instance, one interviewee described the difference between "good smoke" and "bad smoke" as being related to its color and the way it moves through space. When done well, smoke from prescribed fires is seen in wisps of blue and grey, while smoke from wildfires is usually more intense and looks white and grey. The same community member described the bad smoke from wildfires as usually "dirty, dank, and moist," which lays in the valley and does not move, whereas good smoke is perceived as lighter, less moist, and does not linger in the valley. Interviewees also described the taste of smoke as being relevant to their assessment of air quality. One reported, "When you get up in the morning and taste it, it's really bad. It's funky, or close, or there is a lot of it."

Residents in the community are actively making qualitative, spatial, and temporal assessments of air quality, and taking actions for themselves and their community on the basis of these assessments. In discussions with intentional fire practitioners and community residents, several people indicated that the expanded Purple Air network was an additional and helpful source of information to them. For example, a leader in the Karuk Tribe's fire program described how people working on a prescribed fire used real-time feedback from the air sensor network (accessed through a mobile phone) to support hour-to-hour decisions on when to ignite certain portions of an area being treated, aiming to better protect nearby communities while still advancing goals for the treatment area. Another person described how the air sensor network was valued by seasonal fire workers who lived in the area. They use the network to monitor the smoke levels both near large wildfires and in the fire camps to understand and reduce their overall exposure levels and maintain health. Residents described the sensor data as validating what they already knew was bad air quality in their community, while some described the data as striking due to how high the levels were. People also described using the sensor network to identify when and where the air was appropriate for recreation, particularly for children. Some described how having better visibility through the sensor network was helpful for prioritizing outdoor activity, but also noted that some outdoor jobs just needed to be done anyway—in spite of the smoke.

Collection of different kinds of data aids our understanding of smoke across fire events. These data could also help residents in making individual and community health decisions. No single information source provides a complete view of smoke. Peoples' visual and olfactory sensations identify fundamental differences in the qualities of smoke that are not easily measured by low-cost sensors. Still, when the air pollution levels go from moderate to severely unhealthy, people want less subjective measures. The same smoke that appears to be thick and orange, backlit by the sun, could be difficult to perceive hours later. The Purple Air network and other sensors and monitors fill this perception gap and provide more information across time and landscape. Combining visual, olfactory, and taste perception data with numeric information from the sensor network could be valuable for community health and decision making.

Assessing Indoor Air Quality Infrastructure

Many interviewed residents described staying indoors with doors and windows closed as a common strategy used for protecting themselves from poor air quality. Some described using masks (even in pre-pandemic times) for protecting children, the elderly, and other residents with respiratory vulnerabilities. However, outdoor and indoor air quality are related, so the program also focused on analyzing strategies for improving indoor air quality with a focus on homes and community buildings. This portion of the work includes identifying the needs and priorities for air filtration to clean indoor air, and analyzing the utility of indoor air sensors to monitor their performance.

While early discussions included the concept of clean air shelters that could be set up on days with particularly hazardous air, it became clear that these would not be effective or desirable to community members because hazardous air is simply too frequent. It would be disruptive to community life and livelihoods to shelter from frequent events. Therefore, we shifted our focus toward other strategies for clean indoor air such as providing more smaller scale filtration units for residences and community spaces. Presently, researchers are analyzing the relationship between outdoor and indoor air quality in several homes in the Orleans-Somes Bar region, with and without the use of filtration systems. The aim of this portion of the program is to appropriately size air filters for homes in the region given specific building types and high outdoor air pollution levels.

Planning for a Fire-Adapted Energy System

Residents in the Karuk ancestral territory, which is in rural and inland Northern California, face extremely tenuous energy access. During multiple interviews, residents identified power losses a s one of the main issues faced by the community. This problem is exacerbated during wildfire season when utilities increasingly turn off the grid to reduce the likelihood of causing fires. Lack of electricity creates twin problems: 1) not being able to run air filters; and 2) not being able to access information about potentially fast-moving wildfires since communication services are also affected during blackouts.

Analysis of the qualitative and quantitative data identified need for a fire-adapted energy system that would provide electricity to Orleans and the nearby region. The system would enhance the energy resilience of the region, reduce the number of days people would need to live without energy, and allow the Tribe to own and operate its own energy system thereby prioritizing the needs of this specific area of Northern California. Furthermore, having more reliable energy services would allow people to stay connected with the wider world through internet and cell phone communications. The system would also support preservation of culturally significant foods like salmon and acorns: these are harvested at specific times in the year and are stored in freezers vulnerable to power outages. Finally, reliable power would also support air filtration systems and air sensor networks identified as vital for community health.

As part of this study, we assessed the potential of local distributed renewable energy and battery storage to meet Orleans' energy needs today and in the increasingly electrified future. This study included the preliminary design of a solar-powered front-of-the-meter microgrid with a target of $12 million in funding for capital expenditures from Pacific Gas & Electric's Microgrid Implementation Program (an upcoming funding program). Selected results from this study are shown in Table 1.

As shown in Table 1, a $12 million system could provide autonomous power to Orleans for up to three weeks. We also estimated the impacts of electric vehicle and heat pump adoption on days of autonomy. Given 25% adoption, the microgrid could provide up to twelve days of autonomy. At 50% adoption, the microgrid could provide up to 6 days of autonomy, and at 100% adoption, up to four days of autonomy. This type of microgrid could provide improved resilience to community members in Orleans, helping to keep the lights on,

Table 1. Sizing & performance of a $12 million system

Electric Vehicle & Heat Pump Adoption	Solar Size (MWDC)	Battery Power (MW)	Battery Capacity (MWh)	Days of Autonomy
0%	2.0	8.0	31.8	21
25%	2.8	6.1	24.4	12
50%	3.5	4.3	17.0	6
100%	0	12.6	50.2	4

fridges cold, and air purifiers running during wildfires and blackouts.

Working to Bring More Resources to the Tribes in the Region

As a collaboration that has been, from its inception, self-reflexive about the research process and its benefits to entities and people involved, our goals have been explicitly linked to not only research outcomes and academic publications, but also grant writing and fundraising. In other words, bringing resources into Tribal and rural communities has been one of the objectives of the collaboration. The principle of reciprocity is central to the Karuk Tribe's ethos, and guides the process for collaborating with academic institutions, as encapsulated in the Practicing Pikyav document that outlines the process and guidelines by which the Karuk Tribe engages with researchers (*Practicing Pikyav: A Guiding Policy for Collaborative Projects and Research Initiatives with the Karuk Tribe*). The SAFE collaboration has been guided from its inception by the principles outlined in the Practicing Pikyav document. The resources from the SAFE collaborative grant pay for staff time at the Karuk Department of Natural Resources and Blue Lake Rancheria. Furthermore, the research tasks carried out through the collaboration enabled the identification of energy and air quality infrastructure needs in the community. Rather than leaving these identified needs as abstract goals for the future, this collaboration has worked to apply for state and federal funding to enable the identified infrastructures to come to fruition. This practice is part of the partnership's attempt to transcend one-off engagements and create longer term relationships between academic researchers and Native Tribes on shared climate goals.

Thus far, three additional grants have been received by this collaborative group for future work identified by this team's activities. These include a planning grant from the Transformative Climate Communities Program of the Strategic Growth Council of the State of California, with the goal to support and develop community infrastructure in affordable housing, clean energy, air quality, electric transportation, and food security; a grant from the United States Environmental Protection Agency (US EPA) to establish a more extensive air quality monitoring network (for both indoor and outdoor air quality monitoring) across the Klamath River Watershed in partnership with the Blue Lake Rancheria and Karuk Tribe; and a CA100 grant to support the Karuk Tribe's Health and Human Services efforts to develop, implement, and evaluate an updated air filtration program that distributes air filters to an additional 100-200 households. While securing funding for implementation activities is not usually a part of academic research projects, this grant work emerged as a crucial component of properly valuing staff time spent on research at the Karuk Tribe and Blue Lake Rancheria and ensuring that the outcomes of this particular interdisciplinary research partnership benefited all our collaborators in both intellectual and material ways.

Conclusion: Imperfect and Emergent Collaborations can Still Lead to Meaningful Insights

While being far from perfect, our ongoing partnership has illustrated exciting ways that collaborations can bring the critical social sciences together with more technocratic approaches to pressing societal challenges like air pollution, wildfires, energy access, and climate change. While we have experienced numerous challenges of bridging disciplinary divides for the mutual benefit of the intellectual work of the partnership, and several of these challenges are not solvable in the short or medium term, we have also found exciting ways to collaborate and generate insights that are more meaningful and impactful when thought about together.

While many within academia are keen to explore interdisciplinary collaborations, there are very real intellectual and pragmatic challenges in conducting interdisciplinary research. We offer our reflections on these challenges not to discourage our readers from attempting to work across disciplinary divides, but as an honest account of the way the collaborative process has unfolded, and to offer outcomes of collaboration as forged in the struggles of interdisciplinarity.

Most critical social scientists work alone on projects, or on rare occasions with collaborators who are peers, while most STEM-based researchers tend to work in teams with specific tasks allocated in hierarchical ways. Practices of collecting, storing, analyzing, and sharing data are also widely varied between these intellectual traditions. What is considered "data" is fundamentally different and is protected in different ways. For example, "field notes" are closely protected and guarded by ethnographers, while engineering teams consider them in less hallowed terms. The norms of interviewing vary depending on the disciplinary orientation of the researcher. Many critical social scientists prefer semi-structured interviews with questions functioning more as prompts to have a wide-ranging conversation. The grounded, open-ended method of collecting data in this disciplinary context serves an important revelatory function: to use the research to arrive at topics that are most pressing to the field context, rather than to identify them in advance based on the prior beliefs of the researchers. On the other hand, when engineers conduct interviews, many prefer neatly delineated questionnaires with answers kept short and objective. Neither is a universally better method of collecting information; they both allow us to learn different things. These differences extend beyond the open-ended or closed nature of the interview. For instance, while both may be asking relatively direct questions about community experiences of looking at air quality data, an engineer would be interested in whether color coding data in particular ways conveyed information more clearly to viewers, while a critical social scientist would be interested in how the numerical air quality information was assimilated into visual, olfactory, and other sensory perceptions of air quality that residents already relied on. This is just one, albeit illustrative, example of the divergences that emerged through the process of our work together.

Norms around data collection, academic writing, conference presentations, and publishing also vary widely. Critical social scientists are usually expected to publish solo-authored works, rather than writing journal articles with multiple authors, the latter being the norm in many engineering and natural science disciplines. Furthermore, some of the researchers on this collaboration work adjacent to academia, operating within the intellectual and managerial norms of an engineering firm, leading to its own unique advantages and disadvantages of accomplishing research tasks in collaboration with faculty who operate with different incentives and ways of thinking about the labor of intellectual work.

Furthermore, programs that consist of collaborations between university researchers and Tribes can have unique institutional constraints to navigate. Like universities, Tribal governments have their own bureaucratic structures to work within, and academic-Tribal collaborations often need to learn each others' institutional processes and vocabulary of operation in order to successfully navigate working together. In addition, when working on research projects funded by state or federal resources, Tribes could be asked to waive their sovereign immunity as a condition of receiving research funds. This clause acts as a disincentive for Tribes to receive state or federal funds, which limits the research partnerships that Tribes can enter into. On the other hand, university research is funded by state and federal sources, and faculty are incentivized to apply for prestigious state and federal research grants.

Tribal sovereignty is recognition that Tribal governments are independent nations vested with powers. Interactions and contracts with Tribes are with their sovereign governments, through their officials. Sovereign immunity is a protection offered to sovereign entities such that a lawsuit brought against the actions of a state or its officials does not bear adversely upon the citizens served by that state (such as member of a Tribe). It was important to all of us on the research team that we honor Tribal sovereignty, and so a creative administrative workaround was devised by which Cal Poly Humboldt accepted liability for this research collaboration on behalf of our Tribal partners. We also looked for opportunities to address this issue so that the workaround will eventually not be needed. With the platform of a state-funded research project, our team was able to raise the issue of sovereign immunity with staff and leadership at several state agencies through the course of this collaborative work. We hope this will lead to a policy change within state agencies when they are collaborating with Tribes, so that future collaborations with universities on state-funded research are encouraged rather than discouraged by the conditions of state funding.

While our ongoing research partnership has highlighted numerous challenges of carrying out interdisciplinary research in collaboration between researchers in different academic departments, and between universities and Tribal partners, our work has also emphasized some of the benefits of such engaged scholarship. We hope that this short and reflexive article highlights both the joys and frustrations of collaborative and community-engaged scholarship. We believe such partnerships are building blocks for our collective efforts to address contemporary environmental and societal challenges.

References

California Department of Forestry and Fire Protection (CAL FIRE), Stats and Events, accessed here: Retrieved November 14, 2022. https://www.fire.ca.gov/.

Cal Poly Humboldt 2022. “Vision.” Arcata, CA: Cal Poly Humboldt. https://strategicplan.humboldt.edu/#hsu-purpose

Ford, Bonne, Maria Val Martin, S. E. Zelasky, E. V. Fischer, S. C. Anenberg, Colette L. Heald, and J. R. Pierce. 2018. "Future Fire Impacts on Smoke Concentrations, Visibility, and Health in the Contiguous United States." *GeoHealth* 2(8):229-247. doi.org/10.1029/2018GH000144.

Goldman, Mara J., Paul Nadasdy, and Matthew D. Turner, eds. 2011. *Knowing Nature: Conversations at the Intersection of Political Ecology and Science Studies*. Chicago: University of Chicago Press.

Haraway, Donna. 1988. “Situated Knowledges: The Science Question in Feminism and the Privilege of Partial Perspective.” *Feminist studies* 14(3):575-599. doi.org/10.2307/3178066.

Holder, A. 2020. Personal Communication.

Jaffe, D. A., S. M. O’Neill, N. K. Larkin, A. L. Holder, D. L. Peterson, J. E. Halofsky, and A. G. Rappold. 2020. “Wildfire and Prescribed Burning Impacts on Air Quality in the United States.” *Journal of the Air & Waste Management Association* 70(6):583-615. doi.org/10.1080/10962247.2020.1749731.

Liboiron, M. 2021. *Pollution is Colonialism*. Durham, NC: Duke University Press.

Long, J. W., L. W. Tarnay, and M. P. North. 2018. “Aligning Smoke Management with Ecological and Public Health Goals.” *Journal of Forestry* 116(1):76-86. doi.org/10.5849/jof.16-042.

Norgaard, Kari Marie. 2014. "The Politics of Fire and the Social Impacts of Fire Exclusion on the Klamath." *Humboldt Journal of Social Relations* 36:77-101.

Norgaard, Kari Marie. 2019. *Salmon and Acorns Feed Our People: Colonialism, Nature, and Social Action*. New Brunswick, NJ: Rutgers University Press.

Pacific Gas And Electric Company. 2019. “2018 Annual Electric Reliability Report.”

Pyne, Stephen J. 2016. *California: A Fire Survey*. Vol. 2. Tucson, AZ: University of Arizona Press.

Sandoval, Catherine J. K. 2018. “Energy Access Is Energy Justice: The Yurok Tribe’s Trailblazing Work to Close the Native American Reservation Electricity Gap.” In *Energy Justice: US and International Perspectives*. Northampton, MA: Edward Elgar Publishing:166-207.

Subramaniam, Banu, and Angela Willey. 2017. "Science Out of Feminist Theory Part One: Feminism's Sciences." *Catalyst: feminism, theory, technoscience* 3(1).

Tripp, Bill. 2017. “Too Much Fire Elsewhere and Not Enough Here.” In Allison Jolley, ed. “Good Fire on the Ground in 2017: Seven Controlled Burns to Learn From.” Fire Adapted Communities Learning Network. https://fireadaptednetwork.org/7-controlled-burns/.

Tripp, Bill. 2020. “Our Land Was Taken. But We Still Hold the Knowledge of How to Stop Mega-Fires.” *The Guardian*, September 16. https://www.theguardian.com/commentisfree/2020/sep/16/california-wildfires-cultural-burns-indigenous-people.

Western Klamath Restoration Partnership. 2022. “Somes Bar Integrated Fire Management Project.” https://www.wkrp.network/orleanssomes-bar.

HJSR
50TH ANNIVERSARY EDITION

Critical Realignment of Humboldt's "Normal School": Meeting the Changing Landscape of Teacher Education

Heather H. Ballinger <hb481@humboldt.edu> *(Cal Poly Humboldt)*, **Libbi Miller** *(Cal Poly Humboldt)*, **Sara K. Sterner** *(Cal Poly Humboldt)*, **James Woglom** *(Cal Poly Humboldt)*, **Sarah McCue-Green** *(Cal Poly Humboldt)*

Abstract

Our team of teacher educators share our process for reconceptualizing and reaffirming our vision of teacher education as the larger university shifts focus to that of becoming a polytechnic institution. We developed a heuristic to help us articulate our vision based on our commitments and contexts, and discuss the practical implications that this work has in our coursework and clinical practice. Our vision of the polytechnic teacher centers around four commitments: reflective, relational, reconstructive and research-oriented teaching. These commitments are informed by and exist in relationship with broader concepts of positionality, communities of practice, community partners, and polytechnic perspectives.

Keywords: polytechnic, polytechnic teacher, positionality, communities of practice

The Humboldt State Normal School was founded in 1914 in response to a statute mandating the establishment of institutions for the "training and education of teachers...in the art of instructing and governing the public schools" of the state of California (Tierney et al. 2011). In the Spring of 1915, its first graduating class consisted of 15 white women who had successfully met the state's newly adopted teaching credential requirements (Tierney et al. 2011), beginning a storied tradition of preparing teachers to address the demands of the society they would serve (Tanner 1993). From its very inception, teaching-about-teaching has been at the forefront of the pedagogical and curricular aims of what is now referred to as California State Polytechnic University, Humboldt.

Cal Poly Humboldt has gone through considerable changes since its first graduating class received their credentials in 1915, yet it has maintained its ongoing responsibility to prepare teachers to be qualified educators to meet the needs of our state's public schools. This responsibility is informed by a commitment to the children who learn in those schools and the society they will work to coauthor. As the university as a whole considers the paradigmatic shift it will undertake in the coming months and years it is imperative that those of us who are engaged in teacher preparation on this campus critically reflect on the emerging and unprecedented educational needs of the communities that we serve, and determine what more we can do, and how, to meet the continuing challenges facing the communities we serve. In other words, how can we become more innovative in supporting the educational requirements of a diverse population that needs the best and most highly qualified school teachers and administrators?

Presently, the state of California is experiencing a persistent and prolonged teacher shortage that directly impacts the educational experiences of students in our local K-12 schools. Schools see the greatest teacher shortage in special education, math, science, bilingual education, and in rural and low income areas (Carver-Thomas et al. 2021; Lambert 2021), which is reflected in the workforce needs of our regional community. Additional inequities exist as Black, Indigenous, and People of Color are disproportionately represented in the population of students without access to certified teachers throughout the state, which contributes to the persistent inequities and marginalization of Black and Indigenous students and students of color in K-12 classrooms (Cardichon et al. 2020; Sutcher et al. 2016). Teaching and learning in many of these schools results in a non-inclusive curriculum, which often excludes content outside of math and reading. If a school does not have the resources, time or staffing to effectively implement broader content such as arts, science, etc., students lack interactive and developmentally impactful educational experiences (Slouka 2009).

The lack of certified teachers across the state has also impacted educator preparation programs, as there is a need to prepare a large number of teachers at a faster rate. This need often results in an efficiency model of teacher preparation that is measured and defined by the intention of generating the most desirable output at the lowest level of resource expenditure (Bansford et al. 2005). Tasked by the state with addressing the expanding teacher shortage, while at the same time working to meet the curricular and structural standards governed by both the California Department of Education and the state's licensure agency, the Commission for Teacher Credentialing (CTC), teacher education programs find themselves in the double bind (Bateson 1956) of working to recruit and inspire future educators who might expand knowledge and affect societal change, while at the same time providing those future educators a highly refined, efficiency-driven model of what it means to be a credentialed teacher. The state bodies mentioned above often suggest quite expansive curricular expectations through published standards and required exams. They outline and establish the parameters of how teacher preparation must be enacted through accreditation processes. Finally, they define the acceptable output of teacher preparation programs through a handful of teacher preparation assessments. All of these expectations, parameters and assessments serve to constrain burgeoning teachers' conceptions of what might constitute good teaching (Kornfeld 2007). This structure, in collaboration with the effort to affect the largest number of capable, credentialed teachers, in as brief a period of time as possible, and at the lowest possible expense for the system, tends to close off the potentiality of who might become a teacher and what they see as their role in schools and society, as well as narrowing the content of teacher preparation programs. The narrowing of content in teacher preparation can lead to deprofessionalization of the practice of educators, as educators find it necessary to rely on pre-packaged or scripted curriculum that is marketed to guarantee student success without consideration of the learning context or student needs (Schön 1983).

Through our work in teacher preparation, it is apparent that teacher preparation is, in and of itself, an intensive process with multiple factors that could potentially exclude viable candidates and their unique contributions to learning. These factors include, but are not limited to: cost of enrollment and state-based licensure fees, expansive time commitments (often precluding or complicating simultaneous employment), required standardized testing addressing both basic skills and subject matter proficiency, and availability and willingness of partnering schools and mentor teachers. These factors, along with countless others, could well limit the number of potential candidates and the diversity of the candidate pool. In order to begin the process of becoming a California credentialed teacher, a candidate must also hold a bachelor's degree and then have the privileged circumstances to be able to commit to the costs for instruction, examination, bureaucratic encumbrances, and significant time outside of employment (Commission on Teacher Credentialing n.d.). The established state-based mechanisms of credentialing teachers are preclusive by design, winnowing down the field of those who are interested in working in education to the more refined number of those who demonstrate they are capable of efficiently working within the constraints of a system modeled on the efficient dispersal of thoroughly vetted curriculum. A process which devalues ingenuity, creativity, and meeting the needs of varied classrooms and students.

In light of these concerns, critically cognizant teacher educators must actively engage future teachers in a praxis that allows candidates to navigate the labyrinthine systemic expectations of professional teaching. Teacher educators must also reflectively assess their teaching and its effects by researching emerging knowledge and practices, relationally coming to know and work with agents in their shared community, and, finally, working with that community to reconstruct a model of education that responsively meets the needs of that community. Because teacher preparation exists in the gaps, as Ronnie Davey (2013) suggests, between "the school and the academy, between theory and practice, between teaching and research, between the 'real world' and the 'ivory tower' and so on" (p. 2), teacher educators must navigate their already challenging praxis within an interstitial nexus of change.

As such, there is a great opportunity for Humboldt's new polytechnic designation to deepen, expand and shift the field of teacher preparation. The School of Education at Cal Poly Humboldt currently prepares teachers in elementary education, secondary education, and education of exceptional learners. The school also prepares educational leaders in educational administration. We, as teacher educators, maintain a commitment to addressing the workforce needs of California public schools through the preparation of teacher advocates who are prepared and committed to disrupting systemic inequities that exist in our communities. The transition to a polytechnic university provides us with an opportunity to explore, redefine, and grow the focus and purpose of teacher education on the Cal Poly Humboldt campus. These shifts will not necessarily lead to extended time in the program, which would result in additional costs for teacher candidates, but will lead to change in the content, focus and experiences that the program provides, as well as the focus on continued learning and professional engagement after the completion of the teacher preparation program.

The School of Education has consistently and historically engaged students and candidates in experiential, community-based, inquiry-centered learning that is formatively critical of existing structures and institutions that perpetuate inequity and oppression. As we shift to becoming a polytechnic institution, we are now positioned to take a critical look at our systems, including roles and structures, and to make deliberate pedagogical decisions that will define our practices in educator preparation programs.

The Polytechnic Teacher

As we conceive our vision of the polytechnic teacher (PTT), we draw upon four commitments in teacher education that frame our work and seek to inspire and prepare teachers who are committed to teaching practices that are reflective, relational, reconstructive and research-oriented. These four commitments are informed by and exist in relationship with broader concepts of positionality, communities of practice, community partners, and polytechnic perspectives. These founding commitments both exist separately and are inextricably intertwined. The synergy between each of these characteristics of the PTT in practice (figure 1) creates a generative environment for preparing future educators who advocate for socially just educational practices, implement antiracist pedagogies, and work to serve the needs of the communities in which they teach. These characteristics are also deeply rooted in the pledge that our students have been making since 1987 upon their graduation: "I pledge, as an Humboldt community member, to discover the social and environmental consequences of my choices and I will engage in positive change on our campus and in our community" (Cal Poly Humboldt n.d.).

As teacher educators, we engage with pre-service teachers who choose teaching because they have passion for connecting with young people and children. Teacher educators must help candidates sustain their passion when faced with challenges and setbacks and to develop effective habits of mind that will sustain them throughout their teaching career. Teacher preparation programs strike a balance between guiding teachers to support them on their first day in the classroom and to sustain them throughout their career as educators. The concept of sustaining teachers in the field underpins our work to prepare our graduates as "adaptive experts" who emerge from a polytechnic teacher preparation program with "core ideas and broad understanding of teaching and learning that give them traction in their later development" (Bransford et al., 2005 p. 3). A polytechnic educational context confronts the notion that

> To the extent that the world is rapidly changing and that this change will continue to impact educational goals and teaching strategies, prospective teachers need to understands how the natural desire to say, "just tell me what to do" will not serve them optimally for the challenges they will face... efficiency is extremely important; otherwise, we are overwhelmed by novelty. But efficiency is also insufficient if we want to adapt. (Bransford et al., 2005 p. 77)

In our efforts to conceptualize the PTT beyond the constructs of state regulations, requirements, and the efficiency model, we have developed the following framework to articulate a conceptual understanding of our vision of the PTT as a model for educational change and purposeful disruption of ongoing educational systems. The following framework includes conceptions of polytechnic teacher education, as explored in figure 1. We employ this tool as a central framing of tendencies, a set of actions we are pursuing as a guide, and an interconnected framework not structured by a top-down approach. While we articulate this framework here, through implementation it is purposefully designed to maintain flexibilities and adapt to changing context, systems and influences.

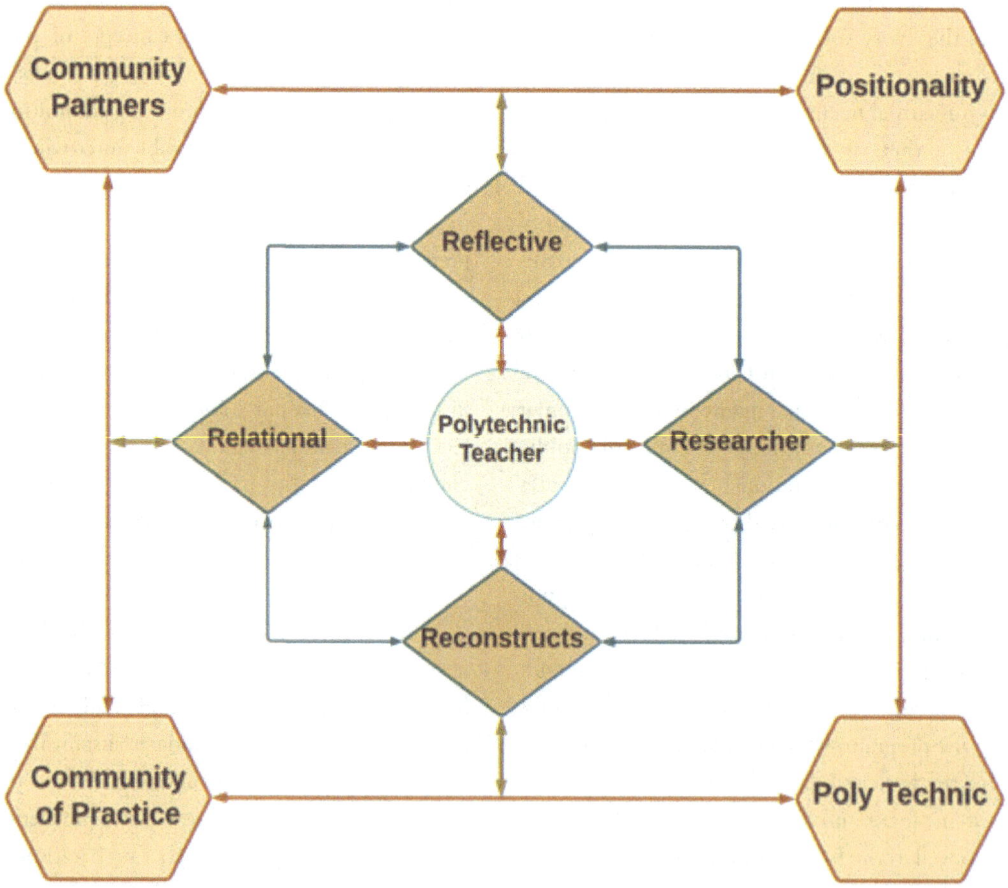

Figure 1: Polytechnic teacher

Reflective

We are in an age of the deprofessionalization of teachers in the field, however teachers continue to be called upon to address, deal with or solve many of society's challenges. Deprofessionalization occurs through top-down, hierarchical, prescriptive curriculum that often defaults to a one-size-fits-all approach to education, as well as in a highly politicized environment that dictates and limits what happens in educator preparation programs and classroom (Milner 2013). The defunding of schools and political movements that devalue education and the expertise of professional educators can shift teaching into a prescriptive model. This removes the reliance on teacher knowledge, experiences and larger contexts of the community. The reflective practice of the PTT model disrupts this deprofessionalization of teachers by centering teacher positionality, leading to considerations and critical examinations of identity, how it may change over time, and how it can be affected by historical and social changes happening around the person and community (Cochran-Smith & Lytle 1999; Kezar & Lester 2010).

Reflective practice is central to shifting from a "how-to", also known as technical-rational, approach in teaching towards increased professionalism with an expanded focus on the positionality of teachers and teacher candidates (Schön 1983). Technical rationality is a framework rooted in the efficiency model and relies on the belief that a technique which is proven to be effective in one setting, application or context will be equally as effective in a similar setting, application or context. The goals of technical rationality include automation, consistency, predictability, and a concerning emphasis on external control. However, the relationships, context, individuality and the human component of the classroom setting is unpredictable. Curriculum designed for one set of students interacts differently with another set of students in a different context at a different moment. The PTT model openly rejects technical rationality in exchange for more thoughtful, reflective practice.

John Dewey (1904) defined reflective thinking as "active, persistent, and careful consideration of any belief or supposed form of knowledge" (p. 9). While Dewey's work provided a foundation for reflection practice, it was Schön's seminal text, *The Reflective Practitioner* (1983), that pushed the teaching field towards a focus on responding to a crisis of confidence. This crisis came out of the increasing reliance on technical rational approaches, such as "step-by-step" curriculum and teaching strategies guaranteed to ensure learning (Schön 1983). Schön contends that reflective practice instead encourages educators to respond spontaneously throughout the learning process and make informed decisions based on the needs of the learners and the ever-changing context. Similarly, Reeder et al. (2006) describes curriculum as a complicated dance in which learning opportunities emerge from the interactions and experiences of the participants, including the instructor as a participant in the group. In their participation, teacher candidates are encouraged to share and reflect on their histories and experiences in order to connect to human experiences through mutual understanding of their own and others' positionalities through focused study and storytelling (Connelly & Clandinin 1990). This kind of reflective practice provides a more conducive environment for the classroom and the teacher's experiences, while avoiding the pitfalls of deprofessionalization.

Relational

The PTT relies on relationality and the process of becoming a relational pedagogue to develop connectedness in and beyond the classroom. Relational theorists place the human relationships between students and teachers at the center of the educational experience and encourage deep questioning of the nature of the relationship and how it impacts the educational experience (Hinsdale 2016). Bateson (1991) defines relationships more broadly and suggests that the world is entirely composed of interconnected relationships, including relationships with broader contexts. He argues that an individual's relationship with these broader contexts provides meaning to the world. Schools and classrooms rely on these relationships amongst individuals, communities, families, curriculum, and content as well as on the larger social and environmental contexts that influence and shape all actions, interactions and experiences.

Reeves & Le Mare (2017) contend that teachers who utilize relational pedagogy are aware of and explicitly focus on the quality of their interactions with students to develop classroom communities that promote academic, social, and emotional growth. Using relational pedagogy as a core framing of a PTT preparation program, the authors center human relations and build community partnerships between Cal Poly Humboldt and P12 schools.

Educational institutions historically rely on hierarchical relationships which utilize a top-down structure to maintain the institution and participant roles. As the PTT begins to explore relational pedagogy, we strive to shift hierarchical (Volk 1995) concepts of relationships to more holistic approaches. The PTT begins to conceptualize holarchical (Volk 1995) structures and relationships for schools and learning communities. In contrast to the pyramid-like hierarchical structure, the holarchy may be visualized as a more circular structure that emphasizes horizontal relationships that are flexible, dynamic and fluid, as the teacher and learner are constantly co-constructing knowledge. In a holarchical structure, participants maintain their roles (Miller 2011; Miller 2016; Petrini 2015). Teachers are still teachers, students are students, etc. Knowledge is constructed collectively through flexible and adaptable relationships. Learning builds upon and sustains the knowledges and experiences of students and their communities (Paris & Alim 2017). This ultimately shifts the role of the polytechnic teachers from a traditional bestower of information to teacher as the facilitator of learning (Freire 2000).

Reconstructive

Social reconstructions are an integral part of these authors' envisioned PTT model. Social reconstruction is a theoretical orientation in education theory that suggests that the practice of teaching is at least partially that of working with students to imagine and sculpt the perpetually emerging social reality in which those students will have authorial agency (Grant & Sleeter 1993) and that we will collaboratively realize and inhabit. This is especially important in light of the institutional theses offered in the Cal Poly Humboldt Prospectus (2022) that suggest that the concepts that will serve to differentiate the Cal Poly Humboldt experience from that of other institutions include a meaningful and intentional focus on social and environmental justice with participation in our local and statewide communities.

The skills and knowledge pertinent to a social reconstructive aim in education must move beyond historical knowledge of and critical attunement to injustice. The PTT commitment includes encouragement towards imaginative and creative practices that, paired with robust anti-racist, socially just and ecologically sustaining curriculum, would provide our teacher candidates and their students the capacities necessary to become ethically righteous agents of change. This commitment stands in stark contrast to multiple trends in teacher education and educational policy in the United States. Legislative actions have been taken in multiple states to constrain or outright ban curricular content, many targeting work that speaks to the convoluted strawman of Critical Race Theory (Bracken et a. 2022). States have been actively and vigorously closing off conceptual connections that might lead students to realizations of injustice and constructive ideation in response to those realizations (Ladson-Billings & Tate 1995). In light of both our university's embedding of ecological and social justice into institutional learning objectives and the national shifts in direct opposition to such commitments and objectives, it is important that we vocally maintain a commitment to social reconstructions in PTT teaching and learning.

As schools continue to grow more racially diverse (Matias & Mackey 2016), teacher educator practices that deeply explore social reconstruction become even more crucial. The commitment to social reconstruction in teacher education is based on the understanding that "continuing business as usual in preservice teacher education [especially in regard to white racialized identities] will only continue to widen the gap between teachers and schools" (Sleeter 2001 p. 96). Equipping PTTs with social reconstruction skills to work with students and communities of practice to dismantle systems of inequities and oppression must be critical to teacher preparation. Because communities of practice and collaboration with our community partners is vital in PTT teaching and learning, it must become a truly synergistic relationship that works collectively to create change.

Research Oriented

The classroom teacher as researcher is not a new concept. Many labels have been used for the type of research teachers conduct in the classroom and at school, including action research (Elliott 1991; Zeichner 1993), practitioner research (Zeichner & Noffke 2001), collaborative inquiry (Bray 2000), critical inquiry (Aaron et al. 2006) and teacher research (Cochran-Smith & Lytle 1993; Cochran-Smith & Lytle 1999; MacLean & Mohr 1999). Commonly, teacher research is a process in which a teacher identifies problems in the context of their schools and classroom and proposes inquiry-based methods to address the identified problems. Through systematic evaluation and data collection, educators observe, analyze, enact changes, and share results to benefit their classrooms and the larger educational community. Teacher research has been acknowledged within the educational research community as a means to promote critical reflection and reform in K-12 settings (Rathgen 2006).

The PTT places research at the forefront of their teaching practice. As inquiry-oriented practitioners, the PTT uses action research methods to improve their teaching, by implementing techniques which are supported through teacher preparation in the development of effective research methods and procedures (Davis et al. 2018). The methods used in inquiry-based research contributes to the development of the PTT's identity and strengthens their current and future practice (Goodnough 2011). Through use of inquiry, the PTT moves away from the transmission model of teaching and develops a transformational focus (Cochran-Smith & Lytle 2009).

The PTT model strives to support ambitious teaching strategies in the classroom through action research and inquiry-based practices. The goals of PTT preparation highlight conceptions of teacher knowledge that Cochran-Smith and Lytle (1999) describe conceptually as knowledge-of-practice. This conception of teacher learning represents a critical constructivist, agent-of-change position that centers the concept that both knowers and knowledge are connected to larger political and social agendas within education. Under this conception, the program, the local community context, and the PTT base their conceptual understanding of learning to teach based on the idea that teachers

> learn by challenging their own assumptions; identifying salient issues of practice; posing problems; studying their own students, classrooms, and schools; constructing and reconstructing curriculum; and taking on roles of leadership and activism in efforts to transform classrooms, schools, and societies. (Cochran-Smith & Lytle p. 278)

Under this conception, being (or becoming) a PTT involves a conjoining of understandings that would ideally occur during inquiry research in which they use their classrooms as research centers to learn more about their students and high-leverage teaching strategies. Bullough and Gitlin (1995) advocate that preservice teachers use research and inquiry to investigate their own positionalities as emerging teachers and develop "personal theories" to think about who they are as teachers and students, particularly with regard to race, class, culture, ethnicity, language, and gender. Advocates for social justice and school change operate within this frame and encourage the PTT to understand themselves as learners and to adopt an inquiry stance to their emerging teaching research.

The PTT's ability and interest in leading research in their field is critical to bringing innovative teaching and knowledge to our community and program. The authors of this paper are interested in contributing to an educator population who are inquiry-oriented so as to foster a cycle of improvement in schools which would ultimately benefit children and adolescents. Research indicates that teachers' engagement in practitioner-oriented action research has a positive effect on their teaching when they are supported in the development of effective research methods and procedures (Davis et al. 2018). With teachers, Goodnough (2011) also finds that the act of conducting research contributes to the development of the teacher's identity and strengthens their future practice. However, even more importantly,

Cochran-Smith and Lytle (2009) argue that this type of inquiry-oriented focus moves educators away from a transmission model of teaching and learning and contributes to a transformational focus.

Implications & Conclusions

With the recent designation of the university as a polytechnic, Cal Poly Humboldt's School of Education is uniquely situated to build a culture within teacher preparation, as well as our local partner schools, that refocuses teacher education to a broader, more reflexive lens. We believe we can address these goals through the visions articulated above, which empowers teachers to take on a professional identity in their practice. The professional identity includes asking critical questions, engaging in inquiry, and taking time to reflect, all while encouraging teachers to consider their own positionality and engage with community partners through communities of practice.

While many teacher educators are interested in and explicitly engage with the concepts explored in the PTT, preparing for a full shift to the framework will require collaboration, engagement and shared-learning opportunities that move away from hierarchical and technical-rational frameworks in teaching and teacher preparation. This type of work can be accomplished through the structured learning opportunities found in professional learning communities and professional development, as well as in unstructured conversations that allow frameworks to emerge, shift and develop in an articulation of existing ideas. The nature of learning in this context is non-linear, multidirectional, and iterative due to the complex, ever-changing and context-driven nature of the PTT framework. The PTT framework is not a one-size-fits-all, unilateral approach to teaching, but an approach that resembles a more complicated conversation amongst stakeholders and participants. The community of practice and community partners must be integrated in this conversation to define the context and provide an essential component of professional learning. In the PTT framework, learning continues long after the teacher preparation program through the development of professionalism and life-long professional learning.

References

Aaron, Jennifer, Eurydice Bouchereau Bauer, Michelle Commeyras, Sharon Dowling Cox, Bren Daniell, Ellen Elrick, Bob Fecho et al. 2006. "No Deposit, No Return: Enriching Literacy Teaching and Learning through Critical Inquiry Pedagogy." *International Reading Association (NJ3).*

Bateson, Gregory. 1991. "Ecology of Mind: The Sacred." *A Sacred Unity. Further Steps to an Ecology of Mind.* New York: Bessie/HarperCollins. (Original work published 1975).

Bateson, Gregory, Don D. Jackson, Jay Haley, and John Weakland. 1956. "Toward a Theory of Schizophrenia." *Behavioral Science* 1(4): 251-264.

Bracken, Kassie, Mark Boyer, Jacey Fortin, Rebecca Lieberman, and Noah Throop. 2022. "What's Actually Being Taught in History Class." *New York Times*, August 17.

Bransford, John, Linda Darling-Hammond, Pamela LePage. 2005. Introduction in L. Darling-Hammond and J. Bransford (Eds.) *Preparing Teachers for a Changing World: What Teachers Should Learn and Be Able To Do.* New York: John Wiley & Sons.

Bray, John N., Joyce Lee, Linda L. Smith, and Lyle Yorks. 2000. *Collaborative Inquiry in Practice: Action, Reflection, and Making Meaning.* Thousand Oaks, California: Sage.

Bullough, Robert V., and Andrew David Gitlin. 1995. *Becoming a Student of Teaching: Methodologies for Exploring Self and School Context.* New York: Garland Publishers.

Cal Poly Humboldt. (n.d.). "Graduation Pledge." September 12, 2022. https://gradpledge.humboldt.edu/grad-pledge-home.

Cardichon, Jessica, Linda Darling-Hammond, Man Yang, Caitlin Scott, Patrick M. Shields, and Dion Burns. 2020. "Inequitable Opportunity to Learn: Student Access to Certified and Experienced Teachers." *Learning Policy Institute.*

Connelly, F. Michael, and D. Jean Clandinin. 1991 "Stories of Experience and Narrative Inquiry," Educational Researcher, 19 (June-July, 1990), 2-14. A version in pp. 121-153 in Edmund C. Short, ed.,

Forms of Curriculum Inquiry. Albany, NY: SUNY Press.

Carver-Thomas, Desiree, Melanie Leung, and Dion Burns. 2021. "California Teachers and COVID-19: How the Pandemic is Impacting the Teacher Workforce." *Learning Policy Institute* (2021).

Cochran-Smith, Marilyn, and Susan Landy Lytle, eds. 1993. *Inside/Outside: Teacher Research and Knowledge*. New York: Teachers College Press.

Cochran-Smith, Marilyn, and Susan L. Lytle. 1999. "The Teacher Research Movement: A Decade Later." *Educational Researcher* 28(7): 15-25.

Cochran-Smith, Marilyn, and Susan L. Lytle. 2015. *Inquiry As Stance: Practitioner Research for the Next Generation*. New York: Teachers College Press.

Commission on Teacher Credentialing. (n.d.). Teaching Credential Requirements. Commission on Teacher Credentialing. https://www.ctc.ca.gov/credentials/req-teaching.

Davis, Janine, Courtney Clayton, and John Broome. 2018. "Thinking Like Researchers: Action Research and its Impact on Novice Teachers' Thinking." *Educational Action Research* 26, (1): 59-74.

Davey, Ronnie. 2013. *The Professional Identity of Teacher Educators: Career on the Cusp?*. Routledge.

Dewey, John. "The Relation of Theory to Practice in Education." *Teachers College Record* 5(6): 9-30.

Elliott, John. 1991. *Action Research for Educational Change*. Maidenhead, Berkshire: Open University Press.

Freire, Paulo. 2000. *Pedagogy of Freedom: Ethics, Democracy, and Civic Courage*. Lanham, Maryland: Rowman & Littlefield Publishers.

Goodnough, Karen. 2011. "Examining the Long-term Impact of Collaborative Action Research on Teacher Identity and Practice: The Perceptions of K–12 Teachers." *Educational Action Research* 19(1): 73-86.

Grant, Carl A., and Christine E. Sleeter. 1993. "Race, Class, Gender, and Disability in the Classroom." *Multicultural Education: Issues and Perspectives*, 7: 59-78.

Hinsdale, Mary Jo. 2016. "Relational Pedagogy." In *Oxford Research Encyclopedia of Education*.

Kezar, Adrianna, and Jaime Lester. 2010. "Breaking the Barriers of Essentialism in Leadership Research: Positionality as a Promising Approach." *Feminist Formations*, 163-185. doi.org/10.1353/nwsa.0.0121.

Kornfeld, John, Karen Grady, Perry M. Marker, and Martha Rapp Ruddell. 2007. "Caught in the Current: A Self-Study of State-Mandated Compliance in a Teacher Education Program." T*eachers College Record*, 109(8), 1902-1930.

Gloria, Ladson-Billings, and William F. Tate IV. 1995. ""Toward a Critical Race Theory of Education." *Teachers College Record* 97(1), 47-68. doi.org/10.1177/016146819509700104.

Lambert, Diana. 2021. "California School Districts Receive Unprecedented Windfall but Lack Teachers to Help Students Catch Up." *EdSource*, June 22, 2021. https://edsource.org/2021/california-school-districts-receive-unprecedented-windfall-but-lack-teachers-to-help-students-catch-up/656711.

MacLean, Marion S., and Marian M. Mohr. 1991. *Teacher-Researchers at Work*. National Writing Project, University of California.

Matias, Cheryl E., and Janiece Mackey. 2016. "Breakin' Down Whiteness in Antiracist Teaching: Introducing Critical Whiteness Pedagogy." *The Urban Review* 48(1): 32-50.

Miller, Elizabeth. 2011. "Ending the Blame Game." *Encounter: Education for Meaning and Social Justice* 24(4), 2-4.

Miller, Libbi R. "Claiming Technology for Democracy and Social Justice: An Approach to Educational Technology." Pp. 19-31 in *Technology for Transformation: Perspectives of Hope in the Digital Age*, edited by L. R. Miller, D. Becker and K. Becker: Information Age Publishing.

Milner IV, H. Richard. 2013. "Policy Reforms and De-Professionalization of Teaching." *National Education Policy Center.*

Paris, Django, and H. Samy Alim, eds. 2017. *Culturally Sustaining Pedagogies: Teaching and Learning for Justice in a Changing World*. New York: Teachers College Press.

Petrini, Rhonda Lee. 2015. "Program evaluation of a tutoring program to prepare disenfranchised students for college." PhD dissertation, College of Education, Walden University.

Rathgen, Elody. 2006. "In The Voice of Teachers: The Promise and Challenge of Participating in Classroom-Based Research for Teachers' Professional Learning." *Teaching and Teacher Education* 22(5): 580-591.

Reeder, Stacy, Darlinda Cassel, Anne M. Reynolds, and M. Jayne Fleener. 2006. "Doing Something Different." Pp. 51-68 in *Curriculum & Teaching Dialogue*, edited by B. S Stern: Information Age Publishing.

Reeves, Jocelyn, and Lucy Le Mare. 2017. "Supporting Teachers in Relational Pedagogy and Social Emotional Education: A Qualitative Exploration." *International Journal of Emotional Education* 9(1): 85-98.

Schön, Donald A. 1983. *The Reflective Practitioner: How Professionals Think In Action*. Routledge.

Sleeter, Christine E. 2001. "Preparing Teachers for Culturally Diverse Schools: Research and the Overwhelming Presence of Whiteness." *Journal of Teacher Education* 52(2): 94-106.

Slouka, Mark. 2009. "When math and science rule the school." *Harper's Magazine* 319: 32-40.

Sutcher, Leib, Linda Darling-Hammond, and Desiree Carver-Thomas. 2016. "A Coming Crisis in Teaching? Teacher Supply, Demand, and Shortages in the US." *Learning Policy Institute.*

Tanner, William R. 1993. *A View from the Hill: A History of Humboldt State University. Archives & Reprint Series (imprint)*. 5. https://digitalcommons.humboldt.edu/reprint/5.

Tierney, Dennis. S., Irving R. Hendrick, Sidney A. Inglis, Richard K. Mastain, Ralph A. Brott, Philip A. Fitch, Linda G. Bond, Lawrence W. Birch. 2011. *A History of Policies and Forces Shaping California Teacher Credentialing*. Commission on Teacher Credentialing. https://www.ctc.ca.gov/docs/default-source/commission/files/ctc-history.pdf?sfvrsn=96050f5_2.

Volk, Tyler. 1995. *Metapatterns Across Space, Time, and Mind*. Columbia University Press.

Zeichner, Kenneth M. 1993. "Action Research: Personal Renewal and Social Reconstruction." *Educational Action Research* 1(2): 199-219.

Zeichner, Kenneth M., and Susan E. Noffke. 2001. "Practitioner Research." *Handbook of Research on Teaching* 4: 298-330.

HJSR
50TH ANNIVERSARY EDITION

What Counts as Literacy in the Polytechnic Hispanic Serving Institution? Culturally Sustaining Frameworks for Writing Assignments, Assessment, and Language Use

Lisa Tremain <ldt142@humboldt.edu> (Cal Poly Humboldt), **Jill Anderson** (Cal Poly Humboldt), **Beth Eschenbach** (Cal Poly Humboldt), **Nicolette Amann** (Cal Poly Humboldt), **Kerry Marsden** (Cal Poly Humboldt)

Abstract

This article theorizes and describes classroom pedagogies that support the development of students' disciplinary literacies through culturally sustaining, socially just approaches. Drawing primarily from the framework of culturally sustaining pedagogies (CSP) (Paris, 2012; Alim, Paris & Wong, 2020), the authors suggest strategies for writing assignment and assessment designs that support students' multilingual and multiliterate ways of knowing. These strategies intentionally invite and integrate students' multiple ways of knowing and being in and outside of the polytechnic HSI. They also ask instructors to decenter the ways that whiteness operates in their curricula and programs. The authors conclude the article by arguing that culturally sustaining approaches must be part of how instructors think about disciplinary literacy development in the HSI polytechnic context.

Keywords: literacy, language, writing assignments, writing assessment, culturally sustaining

W*hat counts as literacy?* This question is a timely and pertinent one for instructors and administrators to ask at Cal Poly Humboldt. As a Hispanic-Serving Institution (HSI) and a newly designated and developing polytechnic campus, what students *do* as writers and readers across courses and in their major programs is imbricated in the Cal Poly Humboldt identification. But literacy is not only school-based, it is also shaped by and reflective of *all of* our identities, our past and present lived experiences of reading, speaking, writing and being. How educators understand and teach toward disciplinary literacy, then, requires them to not only understand students' personal, social, cultural and disciplinary literacies as enmeshed, it also invites them to view students' literacies as dynamic and flexible, as having the potential to enrich learning and each disciplinary community. In other words, we *all* have and display multiple literacies, and these appear in different ways as we navigate the globally networked world and ever-evolving "new technologies, [which have changed] the economies and dynamics of publishing and expand[ed] the possibilities of expression" (Bazerman 2013: 21).

Philosopher Walter Ong articulated writing as a distinct technology that restructures thought (1992). By naming writing as a technology, we can see how this practice works under the broad umbrella of the *polytechnic* institution. The etymological roots of the term polytechnic come from the Greek *polytekhnos*, meaning "skilled in many techniques or arts." While the polytechnic designation for Cal Poly Humboldt has primarily focused on building new STEM programs and hands-on learning opportunities for students in STEM fields, knowledge-making and knowledge-sharing across all disciplines relies on communication, including— and often especially— written communication. While current constructions of the term *polytechnic* at Cal Poly Humboldt tend to appropriately emphasize "hands-on" or applied learning for students, we suggest that the term *polytechnic* on our campus should be re-conceptualized to include all forms of the applied practices of communication: written, spoken, visual, embodied and multilingual. This kind of re-conceptualization can allow campus stakeholders to deeply see and make visible the ways that knowledge is and can be shared in the world.

In this article, we focus on pedagogies for antiracist writing instruction and assessment that bring a *multiliteracies* perspective to disciplinary and polytechnic learning. We present these pedagogies through the framework of Culturally Sustaining Pedagogies (Paris 2012; Alim and Paris, 2017; Alim, Paris, and Wong 2020). Because equity-focused and antiracist approaches to teaching, including culturally sustaining frameworks, are designed to remain fluid and responsive to the contexts in which they are applied, we localize and operationalize the theories and strategies shared in this article in the Cal Poly Humboldt HSI and polytechnic context. In this text, we work as a co-authoring team of cross-disciplinary Cal Poly Humboldt faculty to describe applications of Culturally Sustaining Pedagogies (CSP) to literacy development in the postsecondary classroom. Two individuals of our authoring team, Beth Eschenbach (Environmental Systems Engineering) and Jill Anderson (Kinesiology/Public Health) provide personal reflections that illustrate applications of CSP. Jill and Beth were faculty participants in multi-week faculty learning communities (FLC) facilitated by article co-authors Lisa Tremain, Nicolette Amann, and Kerry Marsden. The curriculum and content of the FLCs focused on antiracist writing assignment and assessment designs with CSP as a common organizing framework. Some best practices for writing instruction and assessment that emerged from across the FLCs are shared later in this text.

But a caution: while this text describes strategies and applications of CSP as part of writing instruction, we do not intend them to be viewed as easy solutions or quick fixes for developing students' literacy. Equity and antiracism don't work in these ways. Instead, the article's purpose is to invite educators to remain committed to the ongoing projects of equity and antiracism in their classrooms, offices, and programs, and to turn toward deeper conversations about how literacy development happens for students in their academic disciplines. We suggest that a first step toward this exploration requires naming and understanding the ways that dominant discourses, namely middle and upper-class white epistemologies, monoculturalism and Standard American English (SAE), have operated in our own and students' educational histories and literacy development. We draw upon Culturally Sustaining Pedagogies (CSP) as *one* approach to decentering whiteness in the writing curriculum and anchoring writing pedagogy and assessment practices to students' diverse ways of knowing and being. Jill's and Beth's reflections (and the theories and pedagogical applications that inform them, as we will share) are intended to illuminate the overarching claim of this text: written communication pedagogies that are attuned and responsive to minoritized students' lived experiences and ways of knowing are essential for Cal Poly Humboldt's vision of a 21st century polytechnic HSI.

We begin the article by drawing theoretical connections between CSP and traditionalized and emergent pedagogies for writing. We then dissect the larger question *What counts as literacy?* into three sub questions: *What counts as writing? What counts as assessment?* and *What counts as language in the discipline?* We offer examples of culturally sustaining approaches that can be applied to writing assignment design, assessment, and language use, and we follow these explorations with Jill's and Beth's reflections on teaching and assessing writing for equity in their own disciplines.

How Did You Learn to Write in the Discipline?

We came to the question *What counts as literacy?* after participating in and facilitating various professional development discussions, workshops and learning communities on equitable writing instruction for faculty across disciplines at Cal Poly Humboldt. Across discussions, we noticed that some instructors found it difficult to break away from long-standing requirements in their teaching of writing. They typically assigned entrenched, long-standing disciplinary genres and prioritized (and punished students for lack of fluency in) Standard American English (SAE), what April Baker-Bell names "White Mainstream English", as part of their writing assessment. While faculty who attended FLC sessions showed dedication to and curiosity about improving student writing, they typically discussed students through deficit frameworks: students "can't write" and do not have the resources necessary to succeed as writers.

But the truth is that students *can* write. They arrive at college with multiple literacies that inform the writing that they do, literacies that are linguistic, cultural, social, and academic. They include, for example, their social group discourses, cross-linguistic and cross-dialect knowledges, and new media literacies (Baca, Hinojosa, and Murphy 2019; Shelton 2019). Yet because typified disciplinary communication practices are reproduced as standards for learning, and because these standards almost exclusively derive from and uphold dominant/white ways of knowing, writing, and speaking, students' rich and various literacies are typically not invited to appear in the most consequential writing that they do for school (see Inoue 2015; Inoue 2016). When students' non-dominant literacies *do* appear, they are often assessed as not meeting standards, requiring correction, and/or punished via low or lowered grades. In this article, we suggest that deep, sustained and reflective professional development that involves collaborative study of and experimentation with antiracist, transparent and culturally-sustaining pedagogical theories can disrupt narratives that describe students' communicative practices in deficit constructs and transform curricular structures to be more inclusive and democratic.

In FLC meetings, we asked instructors to reflect on how *they* learned to write and how they learned to *teach* writing in their disciplines as an entry point into our exploration of the "students can't write" belief. What became clear to us was that instructors' beliefs about writing instruction and writing excellence were deeply rooted in their personal literacy histories and especially in their disciplinary training. Most university faculty's disciplinary training, and specifically their learning and practicing of disciplinary writing, happens in graduate school where they are taught to reproduce, value and uphold entrenched disciplinary genres, particular criteria for "good" writing, and certain beliefs about style and grammar. These genres, criteria and beliefs are not ideologically neutral. They almost exclusively privilege upper-class and middle-class white ways of knowing because they come from racialized academic histories where whiteness is over-represented and over-replicated, where non-white dialects and epistemologies are not viewed as legitimate (Inoue 2016; Baker-Bell 2020; Falconer 2022). When instructors teach students the ways they were taught to write in the discipline, students continue to learn to write (and learn what writing is) according to entrenched/white values. These values are so endemic to U.S. institutions that they are nearly invisible in everyday practice, but they result in gatekeeping, exclusion and/or expectations of assimilation for BIPOC and marginalized students in the academy.

In order to understand how whiteness operates in the post-secondary institution, we can begin by noticing how white institutional presence (WIP) structures it and how WIP works to mask inequity there (Gusa 2010; Falconer 2022)[1]. At Cal Poly Humboldt, collectively and individually, we can reckon with WIP by examining how whiteness is privileged in our classrooms and programs, despite (or in ignorance of) our HSI designation. In her groundbreaking analysis of anti-black linguistic racism, Baker-Bell (2020) articulates this reckoning:

> [E]ducators have to be honest with themselves about the ways they uphold and perpetuate white linguistic hegemony in their classrooms and in their everyday lives. You can't be out here saying that you believe in linguistic diversity at the same time of shutting students down as soon as they open their mouths. You have to be about this life for real for real! You have to be ready and willing to challenge everything you once understood about language and what students need in a language education. You have to be ready for the messiness that comes with this process. You also don't have to do everything by yourself...(100)

This article is an invitation to respond to Baker-Bell's charge: to get messy, to "challenge everything you once understood about language and what students need in a language education." In what follows, we share ways that culturally sustaining approaches to teaching and assessing writing in the disciplines can counter WIP and traditionalized practices that exclude and harm BIPOC and all marginalized students. Culturally Sustaining Pedagogies have the potential to respond to entrenched practices that perpetuate white racial privilege, to honor and integrate students' lived experiences and multiliteracies, *and* to connect them deeply and meaningfully to disciplinary learning.

A Culturally Sustaining Approach to Teaching and Learning

Culturally Sustaining Pedagogies (CSP) expand on decades of research and scholarship about asset-based approaches to education, including Friere's (1970) critical pedagogy, Ladson-Billings's (1995) culturally relevant pedagogy, and Moll, Amanti, Neff and Gonzalez's (1992) funds of knowledge (Paris 2012; Alim and Paris 2017; Alim et al 2020). These and other cultural-asset approaches to pedagogy have long recognized learners' backgrounds, cultures, languages and dialects as strengths that can and should be leveraged in program and classroom designs. Yet a limitation of asset-oriented (and some diversity, equity, and inclusion approaches), as Paris critiques, is that they do not do enough to decenter whiteness. CSP presents a critical framework that shifts educators from beliefs that equity work means teaching marginalized students to perform white, middle-class, hetero and ableist norms, to actively naming and decentering white/monocultural approaches to learning. This shift requires instructors to grapple with the messiness (and richness) that comes with a true integration of diversity and difference in our programs, classrooms, and institutional communities.

CSP requires instructors to "pay close attention to differences and overlaps in categorizations of race, ethnicity, language, and culture, understanding that these identity [and other] markers cannot and *should not be flattened or generalized*" (Gonzalez 2019: 178, our emphasis). CSP also asks educators to name and resist the ways that cultural erasure and exclusion operate in classrooms and academic disciplines due to the overrepresentation of whiteness in them. Shelton (2020) charges instructors to view inclusion as not only a strategy, but as a restructuring of traditionalized approaches for learning and knowledge-making. From her perspective as a Black female scholar, she writes, "To include me is to share the labor of making sense of my intellectual contribution with me, even when (perhaps *especially* when) my ways of knowing, and being, my references and insights are *not familiar or easily accessible* to those of you who are operating out of Western knowledge and value systems" (18, our emphasis). This flipping of the idea of *access*— the recognition that the predominantly white faculty at Cal Poly Humboldt might not themselves easily *access* students' diverse linguistic, literate, and cultural ways of knowing as part of the institutional culture—works to mirror the lack of access marginalized students have historically encoun-

1. We acknowledge that while we refer to the ways that whiteness operates in learning institutions, we are aware of it within the context of intersectionality. We also intend for this term to implicate the ways that other dominant ways of knowing and being intersect and similarly operate in the academy, including male, upper- and middle-class status, heterosexual, cis-gender, and able-bodied ways, to oppress various intersectional and non-dominant ways of knowing, such as BIPOC, working class and economically poor, female, LGBTQIA, and disabled.)

tered when learning disciplinary discourses which privilege whiteness.

CSP confronts white institutional presence by inviting and valuing marginalized students' multiple literacies and different ways of knowing of being in the post-secondary institution, but CSP involves more than the actions of inviting and valuing. To teach through CSP, educators must:

> ...perpetuate and foster—to *sustain*—linguistic, literate, and cultural pluralism as part of schooling for positive social transformation. CSP positions dynamic cultural dexterity as a necessary good, and sees the goal of learning as additive rather than subtractive, as remaining whole rather than framed as broken, as *critically enriching strengths rather than replacing deficits*. Culturally Sustaining Pedagogy exists wherever education sustains the lifeways of communities who have been and continue to be damaged and erased through schooling. (Alim and Paris 2017:1, our emphasis)

When it comes to teaching and assessing writing in each discipline, CSP can work heuristically to reveal students' literacies as more than what they know about academic reading and writing. As Gee's (1989) foundational research on literacy and discourse notes, demonstrations of literacy ask communicators to play different roles at different times. In other words, your literacy appears in different ways depending upon each context where you show it. These contexts ask you to engage in *saying-writing-doing-being-valuing-believing* combinations in order to communicate (526). In this article, while we limit our discussion of CSP-informed approaches for disciplinary literacy development to writing assignment designs, assessment approaches, and language use as a negotiated practice, we emphasize Gee's definition of literacy as saying-writing-doing-being-valuing to remind readers (and ourselves) that the phenomenon of literacy is always sociocultural, deeply embodied and highly contextualized. It actively moves within but also beyond language-based boundaries, including texts like this one.

What do We Mean by "Writing"?

Earlier, we noted via Ong (1992) that writing is a technology that restructures thought. You might draw a map or take a photo to communicate an idea or a message for someone. Probably you have scribbled a reminder note, sent a quick text to a loved one, or added a "like" or a heart to a social media post. You regularly (almost without pause) use multiple and varied language codes—your linguistic knowledge—to inscribe symbols and form text to complete activities. We write across modes (ways to communicate) and media (textual forms or *genres*) to make sense of each day, of tasks and projects, and of our inner and outer worlds (see: Labov 1972; Heath 1983; Prior 2006). In other words, we use codes and symbols everyday as discursive tools; this all counts as *writing*.

Writing is also always circulating. No individual instance of writing exists in a vacuum. When we write, there are various texts connecting to and moving in time and space around each communicative task at hand (Gries and Brooke, 2018; Trimbur 2000). For example, while this article ultimately appears in a scholarly journal as a final edited draft, there were many, many other texts that circulated to form it (e.g., calendar invitations, marginal notes, flowcharts, emails, Post-its, and collaborative electronic documents) and there will be texts that iterate from it (citations, critiques, lesson ideas). The ways that writing circulates across modes and forms illuminate it as a rich social, cognitive, and *applied* practice, an essential working knowledge that illustrates the polytechnic institutional context as a place of applied learning.

As an applied practice and a technology, writing involves ongoing and flexible applications of thinking, composing, and language use in and across modes and media. But how we engage in writing (and as writers) is always informed by the beliefs and values of the communities where we practice it (Gee 1989; Delpit 1992; Delpit, 2013). Perhaps it is obvious to you that the values and practices of writing and communication look and feel different across different communities; we emphasize these differences here to make explicit that writing is never neutral. Discipline-specific and major-based courses, for example, are spaces where values about writing are always present, whether explicit or tacit. As example, rules about "professionalism" (as we discuss later), citation style, concision, "thesis-driven," passive voice, and figure captions are just a few examples of how these values operate differently across disciplines.

As students apprentice to writing in their major programs, they learn what is and isn't valued about communication in the discipline, and they build disciplinary literacy, disciplinary knowledge, and disciplinary identities. As instructors (and experts) from and of the disciplines, we hope that what we teach students *about* them—including what kinds of texts they should learn to write and the thinking they should do in them—*transfers* out of the major program into other contexts, e.g., a profession, public service, or a graduate program. And because we ourselves have learned through and teach in ways that orient learning toward *transfer*, we have formed beliefs and values about particular kinds of texts that students should write. Yet, we can interrogate these beliefs and practices through culturally-sustaining perspectives of teaching and learning, beginning with an assessment of the extent which WIP operates in our disciplines and then working toward inclusive, liberatory approaches to writing assignment designs.

What Counts as a Culturally Sustaining Writing Assignment?

A CSP approach to writing instruction (via Alim et al, 2020) asks instructors to deeply examine disciplinary beliefs and values as captured and reproduced by the texts (*genres*) students are assigned. CSP requires instructors to critique the kinds of disciplinary identities we ask students to "try on" when we ask them to write in particular forms. It also pushes instructors to work with students to analyze who disciplinary texts are written for, for what purposes, whose voices, and ways of knowing are typically represented, and who is excluded.

CSP and genre

Many entrenched disciplinary texts—for example, the constraints of the research paper, the longform essay, or the technical report—come from long-established historical traditions of academic writing, from histories where less than a century ago, it was extremely rare to see any bodies in a university setting other than white males'. Because entrenched traditions of genre and Standard American English (SAE) persist in the academy today, white ways of knowing and thinking have continued to be privileged. As noted above Falconer (2022) names this phenomenon as white institutional practice (WIP), which often appears as neutral communication styles, such as objectivity, linearity, proceduralism, and alphabetic texts, but these styles are always ideological[2]. CSP presents us with an opportunity to examine the genres we assign in our disciplines; it shifts disciplinary teaching toward questions such as: Whose methods and which texts create knowledge? Later in this discussion, for example, Beth reflects on discussions about genre in the field of engineering. She notes that while her understandings and experiments with genre are still emerging, she now intentionally explores the purposes and constraints of the genres of her discipline with students.

Students' new media literacies are an especially rich locale for working with genre in creative and rhetorically powerful ways. Memes, discord servers and TikTok videos are genres that illuminate the ways that students are already sharing information, solving problems, and making decisions as they write in online spaces.

CSP and student choice

Instructors can apply CSP by offering opportunities for student choice in assignment designs, including expanding the kinds of texts that students are invited to write and the subjects that they are invited to write about. There are different methods for thinking about student choice as part of writing assignment design. Applications of CSP might involve students as agents in shaping parts

2. Consider institutions outside of academia, such as law, medicine, and government policy as additional examples of how WIP is similarly embedded in communicative forms across institutions; these kinds of discourses are disguised (by whiteness) as neutral communicative practices, yet they result in social inequity.

of the assignment for themselves and in collaboration with each other, including the kinds of texts they write and the audiences they write for. CSP-informed writing assignments should be relevant to real-world disciplinary problems that reflect students' lived experiences and invite students to summarize, analyze, or create solutions for problems in genres that communicate information to specific (not only academic) audiences. These approaches engage students to critically analyze how they might effectively communicate with different audiences and consider how and why communication about disciplinary knowledge changes across these contexts (Russel et al, 2009).

While dominant (entrenched) disciplinary genres are ones students realistically might need to learn and practice as part of developing a disciplinary identity or as professional preparation, it is also likely they have had little experience writing these kinds of texts, and it is rare that they have written them in their real-world professional or public contexts. A CSP approach to assigning a typical disciplinary genre can open the subject-matter space of the text to include multiple experiences, viewpoints and epistemologies. Students might, for example, write about the people, ideas, movements, or issues they care about within the constraints of the assigned genre, or draw upon their histories, backgrounds, and lived experiences to consider subjects they want to address. Shelton's (2019) teaching of business writing, for example, asks students to craft a data-informed report for a board of directors of a large company to describe the need for lactation spaces and gender-inclusive restrooms. She asks students to explore different rhetorical strategies for composing a policy or report for a white male audience of "bosses" about the needs of a marginalized community—one that students choose. Shelton finds that "undergraduate students who are learning to communicate with the specialized technical genres and rhetorical conventions of their disciplines are often operating under a logic that suggests to them that difference—including differences among bodies—should be either neutralized or commodified. This epistemology forecloses the possibility of difference as a way to creating…more effective communication" (18). Shelton's point is that *difference* can be a critical and generative framework for fostering disciplinary literacy development. CSP illuminates difference (from the dominant discourse) as a distinct shaping component for how students might be invited to write in response to disciplinary assignments.

What Counts as Culturally Sustaining Writing Assessment?

Traditionalized approaches to writing assessment have also structured students' understandings about writers, writing and learning. As we've noted, these approaches privilege and perpetuate white ways of knowing. Inoue's (2015) book *Antiracist Writing Assessment Ecologies: Teaching Writing for a Socially Just Future* tackles and resists traditionalized assessment and grading designs (e.g., disaggregated rubrics, decontextualized criteria, holistic letter grades) to focus on students' time and labor toward their learning goals. Inoue analyzes commonly used writing rubrics and grading schemes to show the ways these privilege SAE and mastery of normalized (but not neutral) standards for writing. Inoue argues that these approaches harm marginalized students because they require them to assimilate white-dominant literacies in order to be successful in school: they are punished with low(er) grades or excluded from advancing when they do not assimilate white literacies. When marginalized students are evaluated according to this privileging, these structures also reproduce hierarchies that uphold white supremacy. Inoue argues for reimagined writing assessment designs that facilitate environments where learners improve via feedback rather than evaluative marks, where they can focus on ideas, as opposed to mistakes. While this article does not cover Inoue's theories and practices of labor-based grading in detail, we share two culturally sustaining assessment practices for writing assessment that evolved from his work.

Like the normalized and entrenched genres of the disciplines and professions, grading rubrics, too, have material and social effects. Qualitative and subjective terminology in rubric categories, such as *clear*, *sophisticated*, *organized*, and *sufficient*, are those we might see in a typified disaggregated rubric. But these terms homogenize beliefs and values about writing and make assumptions about how students should move "toward the alignment of knowledge"—or code for practicing and performing whiteness (Falconer quoting McMullin, 40). Not only do these terms reproduce white beliefs and values, they also result in gatekeeping, where some students meet the criteria because they have had experience with, access to and practice in the discourse, while other students have not.

Stripping qualified terms from writing criteria to the *minimum essential* engenders another culturally sustaining approach to writing assessment design. This action begins by distilling the minimum essential criteria for proficient writing and removing qualified or arbitrary terms from these criteria. For example, we might shift a criterion for *analysis* from "*clearly* and *accurately* evaluates conceptual and factual claims" to "evaluates conceptual and factual claims." This kind of intentional rhetorical shift for writing criteria removes any interpretive and opaque terms from evaluative criteria. A minimum essential approach to criteria-building strives toward achievable and transparent learning goals. It allows students to justify how they define and practice assessment and evaluation.

The minimum essential design also supports alternatives to the highly subjective process of assigning grades to writing. It gives students room to write from and beyond the criteria in ways that can be creative and innovative. Beth's reflection identifies her own shifts to move toward ungraded and formative feedback on student writing that helps them to move toward or creatively beyond minimum standards. This approach, as she articulates below, results in more students successfully developing conceptual and disciplinary understandings.

As a companion to minimum essential designs for writing assignment criteria, instructors can also replace scaled structures and point values in their grading rubrics. One-point rubrics, for example, simplify evaluative categories to *met* or *not yet met* as opposed to *below proficiency*, *proficiency* and *exceeds proficiency* evaluations in traditionalized rubrics (See Figure 1). The one-point design resists and transforms letter grade and 6-point rubric hierarchies to focus on (ideally formative) feedback aligned to each criteria for success, and it can support students to see their writing in the discipline as developing. Students can, for example, notice patterns of "not yet" across writing criteria and can revise their texts to address "not yet" evaluations. Even when students consistently "meet" all criteria in a one-point rubric, written feedback from their peers and instructor about each criterion can identify next steps for growth. An added benefit: When these two strategies are combined (e.g. using

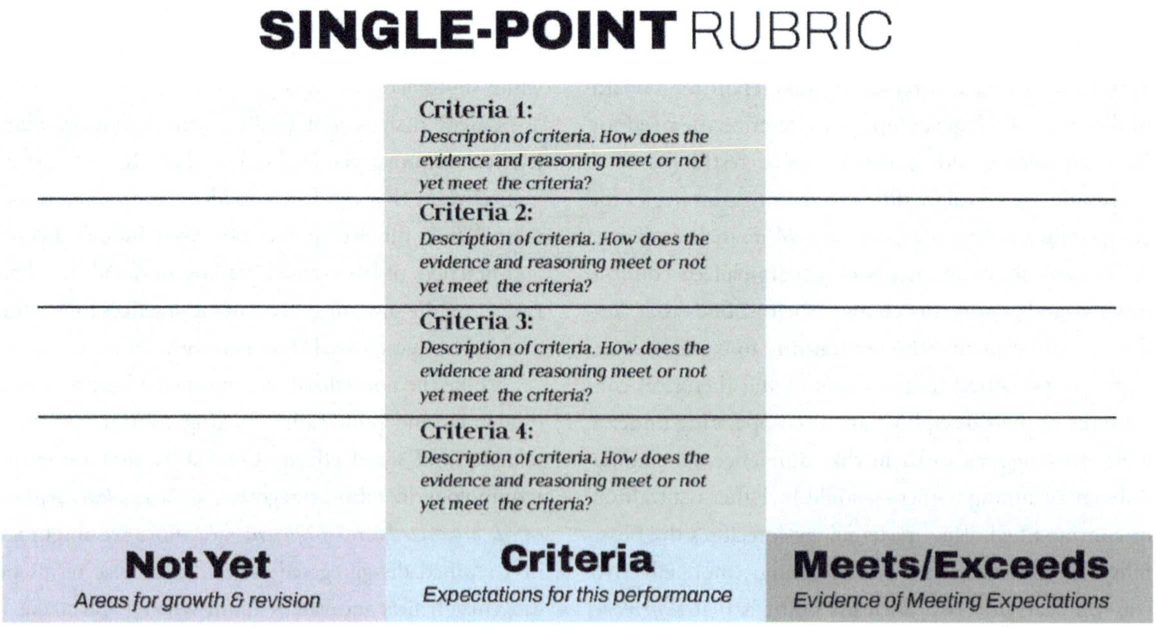

Figure 1: Single-point rubric template, adapted from Jennifer Gonzalez (2015).

one-point rubrics to assess minimum essential criteria), they are easier for instructors, student writers and their peers to measure. Yet instructors must be sure that their assessment designs are grounded in equity, not efficiency. When instructors experiment with new writing assessment designs, CSP requires them to commit to exploring ideological questions related to teaching, learning, and grading, including the purposes and material effects of evaluation on students in relation to each assignment.

Echoing the discussion of *choice* as part of writing assignment design in the previous section, students are also more engaged in their learning when they have choices in how they are assessed. Minimum criteria and one-point approaches are only part of what it means to design holistic approaches to writing assessment. These designs should include students' self- and/or peer-assessments as valid and valued parts of their responses to each assignment. Jill's reflections later in this text describes an innovative approach to peer feedback in her classroom. These approaches to assessment emphasize learning over evaluation and, when used transparently and consistently, they emphasize a culture of feedback and collaboration where everyone's learning, including different kinds of linguistic knowledge, matters.

What Counts as Language?

Each major program on the Cal Poly Humboldt campus is a distinct discourse community that uses language to negotiate, shape, communicate, and challenge its ever-evolving values and practices. Discourse communities and the language in them are fluid; they change across time and as participants move in and out of them, though some discourse communities are more fluid than others (e.g., consider again the fairly static discourses of law, medicine, and government policy). While each discipline uses a specialized language (a lexis), discourses can and do change over time and through negotiated participation (see: Gee 2004).

Now pause to think about how many different groups (or discourse communities) you belong to. How have you—consciously or not—learned and practiced the expectations of language in each of them? Teams, clubs, social groups, workplaces, online groups, gaming communities, and major programs are all examples of discourse communities. The discourses that are used in and across these groups change through members' negotiation of language over time. Humans tend to use various translingual, code-switching and code-meshing practices as part of navigating the changes across discourses. Understanding that language is always negotiated through participation can help instructors critically analyze the disciplinary lexis through key questions: Who has language power? Who polices language? What styles of language dominate and what identities are represented through language? How do I negotiate with students about language? And: where does whiteness and linguistic racism appear in my approaches to teaching disciplinary discourse?

Professionalism, for example, as a concept that students are often expected to perform in writing in their major programs, helps us explore these queries. For many students from marginalized identities, expectations for professionalism in written communication can be coded as the expectation to perform whiteness; this means they need to engage in code-switching in order to be successful as a writer. When code-switching, language users leave behind one communicative code and switch to another. In our disciplines, this means that some students believe they need to (or they are required to) abandon a code, such as their heritage dialect or language, in order to assimilate and perform a white privileging disciplinary discourse, which is typically viewed as what it means to sound and be "professional." Instructors can work with students to explore the extent to which *professionalism* (or the professions where the discipline is applied) has shifted over time in the field or how it changes depending on geographic or community contexts. Later in this article, Jill discusses expectations for "professional language" in the field of health promotion and her explorations to design assignments that allow students to convey concepts of the profession through a variety of language styles and approaches.

As an alternative to the ways that code-switching promotes assimilationist practices that reproduce white ways of knowing, instructors can apply CSP frameworks to sustain students' multilingual and multi-dialectal knowledges by inviting *translingual* and *code-meshing* orientations into their teaching of writing (see: Canagarajah 2013; Canagarajah 2016). Hybridized cross-cultural languages like Spanglish or social languages like internet speak are examples of how translanguaging and code-meshing work in creative and strategic ways to support meaningful communication (e.g., "*Watcha le*" or "*I googled it*" or "*It was gr8 to see you.*"), but these are rarely valued in the written work that students do for school-based assignments. Acknowledging the flexibility and creativity of translanguaging and code-meshing in the writing students are asked to do, however, can yield new ideas, structures, and insights *about* the discipline and disciplinary work itself. Instructional and assessment designs that maintain flexibility about how students use language give them permission to innovate and creatively communicate to meet expectations of disciplinary discourse communities.

But teaching and assessing language in flexible ways does not mean there are no "rules" for language. Rather, as language scholars Lu and Horner (2015) illustrate in this discussion about translingual practices in the classroom, it means shifting dispositions toward and about language:

> [T]aking a translingual approach does not prescribe the forms of writing that students are to produce. Instead, it calls on students (and their teachers) to develop specific dispositions toward languages, language users, contexts and consequences of language use. By recognizing writers' agency in and responsibility for all their language productions, whether these seem to reproduce standardized forms of writing or deviate from them, [a translingual approach] is applicable and of benefit to all students (and all writers), including those deemed mainstream monologues and those deemed multilinguals. (28)

Understanding our own and others' agency as language users is key to understanding how and to what degree these practices can work to address linguistic racism. Instructors might, for example, examine how whiteness and/or linguistic racism operates in the field. This kind of work means instructors—and students—must understand the stakes at hand: translanguaging and code-meshing in some communities or contexts might be extremely beneficial; they might help writers communicate or create new meanings and knowledge, and students can link linguistic flexibility to how problems are solved or addressed. In other contexts, translanguaging and code-meshing may be risky; by using them, a writer or speaker might fail to meet the discourse community's standards, which could mean losing or not getting hired for a job, low grades, marginalization, or exclusion. We can work with students to explore the benefits and risks of translanguaging and code-meshing in our disciplines and support their agency in deciding whether and how to make choices around languages use. This requires instructors to understand students' use of translanguaging and code-meshing as flexible, dynamic, and creative ways to communicate rather than *right or wrong* ways to communicate.

Integrating students translingual and code-meshing affordances into teaching does *not* mean students shouldn't pay attention to (and learn and practice) the codes and expectations of a disciplinary discourse; rather, it means that instructors can help students develop critical awareness of disciplinary norms and expectations and how these work in asymmetrical relations of power within the discipline and institution. As Canagarjah notes, "However unfair and limited they may be, [disciplinary] norms and ideologies have to be taken seriously. Social and educational success means engaging with these norms, though this doesn't mean uncritical acceptance or conformity" (2013: 9). So, while instructors may have strong justifications for why students should practice and develop fluency in a particular language style in order to meet disciplinary expectations, they must remain critically conscious—and help students develop critical consciousness—of how these norms operate.

There are various practices that support culturally sustaining approaches to language use in writing instruction and assessment, such as:

- Ask students about their language backgrounds. Discuss the different ways they will use language early in the term to situate discussions about disciplinary discourse.

- Discuss disciplinary discourse as a *fluency*. Acknowledge how WIP informs the rules of fluency.
- Provide students with both low- and high-stakes opportunities to write in their own voices and through their linguistic experiences.
- Explicitly invite students to consider when and how their linguistic and dialectical practices can be purposefully applied to their writing.
- Unpack subjective and racialized terms like *professionalism* with students and invite them to redefine these terms in ways that feel safe and accessible to them.
- Don't view language as a problem that needs fixing. Recognize that *error* itself is culturally constructed; it is where one or more members of the communicative act refuse to participate in negotiation.
- Avoid penalizing students for perceived language errors. Point to patterns in language use to help students determine whether and how they will move toward disciplinary or language fluency.
- Engage students to review their language choices through inquiry: "This word/sentence confused me because ___ . Can you tell me about this choice?"
- Be willing to let what confuses *you* to stand as a purposeful rhetorical or linguistic choice of the writer. This approach models language as negotiation.
- View and celebrate language differences as the norm.

This limited list is intended to call for a larger shift: programs and individual faculty at Cal Poly Humboldt must account for the ways that white language and expectations of its assimilation work against the HSI designation and specifically against Chicanx/Latinx students' success and the success of all marginalized students. Enriquez-Loya and Leon (2017) argue that HSIs cannot only consider "curricular topics [in their HSI design] but [they must also consider] the practices that reflect our Chicanx/Latinx beliefs and rhetoric, as well as learning outcomes and assessments that utilize our HSI identity as part of our design methodology" (214). To make our HSI identity visible to students, Cal Poly Humboldt instructors can invite translingual/code-meshing practices for writing into their instructional designs, and they can practice feedback and assessment practices that inhabit language as negotiated practice. These approaches require faculty and administrators at Cal Poly Humboldt to maintain a critical consciousness about how students learn, including acknowledging and working to transform the ways that racism impacts literacy, language and writing in postsecondary institutions.

Reflections from Two Cal Poly Humboldt Faculty on Teaching and Assessing Writing

Though we have separated the broad question *What counts as literacy?* into three questions and sections above, developing disciplinary literacy cannot be separated so neatly. Literacy development requires learners to engage in nearly simultaneous and integrated practices of reading, writing, speaking, listening and being. In order to maintain a critical consciousness toward writing instruction and assessment, instructors must experiment with, reflect on, and adjust instruction in collaboration with their students and colleagues. Instructors also need models that show the literacy of the discipline in action. In this section, two reflections as written by two Cal Poly Humboldt faculty, Jill Anderson (Kinesiology/Public Health) and Beth Eschenbach (Environmental Resources Engineering), consider literacy development (their own and their students') as integrated actions of reading, writing, speaking, listening and being. Both Jill and Beth attended professional development workshops and/or multi-week faculty learning communities on equitable approaches to writing instruction at Cal Poly Humboldt facilitated by article co-authors Lisa, Nicolette and Kerry. Jill and Beth reflect on how they learned to write and teach writing in their disciplines, and how they are currently thinking about and teaching language, writing and assessment as instructors in their fields.

Jill's reflection

My journey of learning to write within my discipline was sink or swim. In courses, I predominantly wrote in response to high stakes assignments. Occasionally, an instructor required peer feedback or an outline, but the writing I submitted was expected to be fully formed, final versions. There was not much emphasis on revision. My papers were returned to me marked up with the instructor's colored pen of choice. Maybe we figured out what went wrong before we submitted the next paper and maybe we didn't. I would visit a professor's office hours to ask questions, of course, and they were happy to talk, but revision and review in advance of a "final" were not built into the writing process. Then there was the added expectation to write for peer-reviewed publications and the joy (sarcasm) of hearing from anonymous yet brutal reviewers (aka Reviewer 3). In my field, where qualitative, personalized work is not always valued, academic writing often means I need to adopt a robot voice: concise as possible and following a strict set of discoursal rules. Some of these are clearly defined (like English language mechanics) and others are still unwritten and unnamed, passed down in the culture of the field. If I'm reflecting honestly, I have internalized a (probably self-imposed) pressure to not make mistakes, to be deemed 'good enough' on the first try, and to essentially always write with the voice of Reviewer 3 in my head.

I also never received explicit or meaningful training about how to teach writing; I learned through modeling. I don't mean to suggest I had bad models, simply that any training I had focused on rigid rules as well as on meeting long-established tenants of the discipline. I used these experiences (initially) to help me teach writing at Cal Poly Humboldt. How I "learned" to teach writing was to provide examples, but mostly to just throw the assignment out there, see what comes back, and then mark all over each paper. This process for teaching writing meant that I assumed students would read my comments and corrections, yet I didn't yet know how to structure purposeful engagement with writing as part of an iterative process.

After attending professional development on equitable approaches to teaching writing, I learned, generally, that writing is connected to lived experience. Writing instruction can provide opportunities to invite students in, to help them find connections to the field. I also learned that writing—as an expression of the self—is not simply reserved for "creative" writing situations but can be tapped in all its forms. My developing understandings have led me to ask questions about how writing works in my field, and I have noticed unnecessary boundaries for writing that keep students from accessing connection points to the discipline. My questions have become: how do I foster a balance between meeting the essentials of the discipline, of being able to speak the field's language, alongside supporting students to find connection points? I have asked myself: Why not start with *connection* and then scaffold in disciplinary rules and mechanics? Why centralize mechanics so much that students' connections to the discipline get lost? I now think that putting "rules" first creates a risk where students, as future leaders in the field, can no longer see a place for them in it, where strict structures of writing are, intentionally or not, holding a door open for people to *leave* a field that needs them. My takeaways from professional development in writing have led to fundamental shifts in how I view it. I now see writing holistically. These meetings were, honestly, the first times I had ever really thought about the potential of writing as an inclusive practice.

As a kinesiology and public health professional, I have lived in the world of academic-based writing and writing for public audiences (e.g. infographics, health promotion program materials, etc.). While these formats, intellectually, have always *counted* as writing, students' work to create them has *not* been valued at the same level as "academic" texts. I'd like to see how assessment or evaluation of all of the texts students create in these programs might change to respond to students' labor and efforts. At this point, I am also viewing writing as not simply as an end product but as a process for thinking. I more frequently use writing to engage students in thought exploration and finding connections to content, for engaging students in collaborative critical thinking, and for reflection on their way toward a final product.

In the past, I also required rough and final drafts as components of writing assignments, including peer review (e.g. students would swap papers and provide written feedback to each other). These processes are good, but I learned in the FLC that the writing process and writing assessment can be much more thoughtful and dynamic. In the FLC, for example, we explored a peer review process where students pose questions to the group that they find themselves grappling with while writing. In implementing this approach in my classes, students reported that this process was much more meaningful for them than other peer-review processes we had tried, and that they took away not only editing suggestions but also different viewpoints to consider and critical feedback for moving forward.

These and other new approaches toward writing instruction have shifted how I think about grading and assessment, and how I approach writing as an engagement tool. In the health promotion field, we need folks to be able to create materials that will reach and connect with different groups of people in the general population. At the same time, the field also requires folks to be able to write in formal language as expected in grant or program proposals. Students need to experiment with what it means to write for these different audiences, but a strict separation of language approaches for different audiences may not always be so necessary. While the use of "professional language" is an integral part of working in my field, how assignments invite students to convey professional concepts can happen through multiple language styles and approaches. I now check my lens when I review students' work; I step back and ask myself which kinds of corrections and feedback will best support each student's development as writer and a learner, and whether or not any of my feedback might be unconsciously coming from the strict writing style ingrained in me during my own education. Separating out these elements has created more room for me to engage with students about their writing and their connection to the content, the field, and their peers.

Moving forward, I hope to learn more about writing across the curriculum and would like to see how these processes can be explored within programs. This might be especially tricky when faculty have different areas of expertise within a program and may have different values and constructions of what they think is important for writing success in the major.

Beth's reflection

I learned most about how to write in my discipline during graduate school. Before graduate school, my father commented on his surprise that my ability to communicate in writing did not live up to my verbal ability. Early in my graduate program, I learned that my advisor also did not think I wrote very well. So I did not have a lot of confidence in my writing abilities in graduate school—at least at the beginning.

Many of my advisor's courses had large reading packets that had many of her papers in them. I learned a lot from reading her writing, as well as from other papers, about what content and tone is expected in a journal article. Reading my advisor's papers impressed upon me, for example, how important it is in the introduction and the literature review to carve out a place where one's work contributes to the literature, to show that no one else has made the same contribution and why the contribution is important. My writing was pretty rough until I developed an editor's voice about my own work, which happened through editing other people's writing. Once I got better at explaining why something did not make sense to me, then I became more skilled at editing my own work.

I do not remember teaching writing before I was hired at Humboldt in January of 1995 as an assistant professor of engineering. The first course where I taught technical writing was for first-semester, first-year students in ENGR 115: Introduction to Design. The students wrote large reports in groups which included five major sections submitted over the course of the semester. Students received both formative and summative assessments for their reports. At the end of the project, their "clients" received their final reports and I graded those. I was provided a pre-existing departmental rubric for grading, but I also developed a set of codes for helping students improve their technical writing skills.

Given my own challenges with learning technical writing, when I first started to teach it, I looked hard for ways to help them. I found a technical writing text that had helpful hints which I shared with students. I also shared and continue to share "The Science of Scientific Writing" (Gopen and Swan 1990), which I used in graduate school (Some students have exclaimed, "Why didn't you share this with me earlier!").

In our department, we had developed a framework for writing during the late 1990s when our program learned—from alumni surveys—that we were not teaching technical writing well. We learned that students received lots of feedback on *final* drafts, but they were not provided an opportunity to review feedback and implement it in advance of those final drafts. Some engineering students study engineering partly because they do *not* like to write. I often share with them that "you can have the greatest idea, but if no one else can understand your idea, it cannot be implemented." Though I was already a believer in peer review and I understood that humans share and create knowledge through writing, the FLC sessions reinforced the value of feedback, peer review, teaching writing as knowledge creation.

Genre knowledge is a concept that I am still in the process of understanding. Through my own emerging understandings of genre, I feel I have better structures to explain to my students the reasons for learning a particular genre and why it will serve them to do so. The concept of genre also helps me explain the distinct parts of a particular text. The importance of providing example documents has also been reinforced for me. I was providing examples before, but now have deeper understandings of why examples are useful to students and how to make them useful.

I am still grappling with how to dismantle white supremacy within the writing in my discipline. I am not yet sure what that looks like. Simple ideas come to mind: allowing documents to be written in multiple languages or multiple versions of English. I think I am just beginning to think about "what counts as language" in ENGR. I still suggest that students revise their work to use different words that are more precise or concise. This means I am still pushing the values of concision and precision in technical writing, which also means I am still grappling with how to address and transform white supremacy in writing instruction.

The idea of grading for labor and not on final "quality" is another idea that I am still chewing on (I put "quality" in quotes, because it could be that "quality" is infused with white supremacist notions, or that a student comes into my class writing well, and no "quality" is added by my instruction or assignments). One of our readings in the FLC was about "ungrading" (Heikkinen 2022). The article describes how an instructor's grading system changed to center what students learned as writers in his class, and not on how well they could write before they came into it. The students earned higher grades overall after he made this shift, but he also felt he had honestly assessed students on what they had learned during the term. I would like to get better at assessing students on what they have learned as writers in my classes, and not so much on how well they write in the first place. I now provide ungraded feedback on drafts and encourage students to reach a minimum standard on the first draft. Then, I work with them to extend ideas on each subsequent draft. As a result, I am seeing students earn higher grades in the end.

I am interested in the most effective way to help students learn to improve their writing. To me "effective" means that I use strategies for teaching writing that are both efficient for me as an instructor but also support students' growth as writers. So I wonder: What does the research say? What types of developmental models are there for technical writing development? Do they exist? Are they worth considering?

Reframing What Counts: Writing Instruction in the Culturally Sustaining HSI

Perhaps as you've read this text and these reflections, you've explored your own memories about how you learned to write or your multiple and varied literacies and how they developed. You likely have beliefs about what literacy *is* (in your discipline and otherwise) and how humans become literate over time and across situations. These ideas and beliefs come from schooling, reading, professional experiences, discussions, research, social groups, family, and the systems that conscribe or inform them. In Jill and Beth's reflections, we see the ways that complex webs regarding expectations, genres, affordances, dispositions, and beliefs and values about writing take shape across their experiences as writers and teachers of writing. We see Jill and Beth thinking hard about writing and literacy development in their disciplines. Their reflections tell us how they are thinking about what counts as *literacy in their disciplines*. What is less visible in these reflections—though they are there—is how these instructors' disciplines have privileged (and still privilege) white ways of knowing. In Jill's reflection, she discusses the pressure of meeting the requirement of polished, perfect texts in her graduate program; and, in Beth's, she interrogates the concept of writing "quality" and how that has been constructed in her field and, therefore, in the assignments she's designed. These reflections illustrate Jill's and Beth's thoughts about teaching writing, but they are also about their identities as writers. They illuminate that our *literacies* are personal. They are also *cultural*, *social* and *epistemological*.

CSP provides instructors with a framework for understanding literacy as a cultural, social, and epistemological phenomenon. CSP-informed teaching makes visible the idea that *literacy* involves more than just writing, that it is infinitely more dynamic than one (white, dominant) way of knowing and communicating.

To close this article, we gesture toward hope: hope that readers feel connected to and challenged by the ideas and practices posed in this article; hope that readers seek out and experiment with ways to center their students' multilingual and multiliterate ways of knowing in their writing and their lives; hope that instructors collaborate with their students and their colleagues about their successes and (creative) failures toward equitable writing instruction, and hope that all students learn in culturally sustaining ways. So, some questions for all readers (and not only instructors): How do your beliefs about literacy play out in the ways you think about teaching and disciplinary learning in the polytechnic HSI? How can instructors support students' diverse and varied literacies in ways that sustain them? How can instructors resist and critically address white institutional presence in writing instruction and assessment? How can instructors center students' multiliteracies?

At Cal Poly Humboldt, these questions can help instructor critically and meaningfully respond to institutional mandates related to educating and graduating a diverse body of students and the imperative to create new polytechnic programs. Enriquez-Loya and Leon challenge "all individuals [at the HSI] to break molds and build alliances through shared knowledge and to create community within academic and localized spaces" (2017: 214). Pedagogical frameworks like CSP (and other) equitable approaches to language and writing support knowledge-sharing and community-building across perspectives and experiences, but they cannot by themselves dismantle the racialized structure of the institution. In order to stay connected to the ongoing project of antiracism, instructors must see literacy and language use in all of their negotiated complexities and in ways that move beyond the classroom. Historic and recurrent English-only movements, attacks against public libraries, book and curriculum bans, and ongoing efforts toward voter suppression illuminate critically conscious understandings of literacy as central to human rights. The politics of writing are not only linguistic.

References

Alim, H. Samy and Django Paris (Eds.). 2017. *Culturally Sustaining Pedagogies: Teaching and Learning for Justice in Changing World*. New York: Teachers College Press.

Alim, H. Samy, Django Paris, and Casey Phillip Wong. 2020. "Culturally Sustaining pedagogy: A Critical Framework for Centering Communities." Pp. 261 – 279. *In Handbook of the Cultural Foundations of Learning*, edited by Na'ilah Suad Nasir, Carol Lee, Roy Pea and Maxine McKinney de Royston. Routledge.

Baca, Isabel, Yndalecio Isaac Hinojosa, and Susan Wolff Murphy (Eds.). 2019. *Bordered Writers: Latinx Identities and Literacy Practices at Hispanic-Serving Institutions*. New York: SUNY Press.

Baker-Bell, April. 2020. *Linguistic Justice: Black Language, Literacy, Identity, and Pedagogy*. Routledge.

Bazerman, Charles. 2013. Global and Local Communicative Networks and Implications for Literacy. Pp.

13—25 in *Literacy as Translingual Practice*, edited by Suresh Canagarjah. Routledge.

Canagarajah, Suresh. 2013. "Introduction." Pp. 1-10 in *Literacy as Translingual Practice: Between Communities and Classrooms*. Routledge.

Canagarajah, Suresh. 2016. Translingual Writing and Teacher Development in Composition. *College English*. 78 (3): 265-273.

Delpit, Lisa. 1992. Education in a Multicultural Society: Our Future's Greatest Challenge. T*he Journal of Negro Education*. 61 (3): 237-249.

Delpit, Lisa. 2013. The Politics of Teaching Literate Discourse. Pp. 285 – 295 in *Freedom's Plow: Teaching in the Multicultural Classroom*, edited by Theresa Perry and James W. Fraser. Routledge.

Enríquez-Loya, Aidé and Kendall Leon. 2017. Chicanx/ Latinx Rhetorics as Methodology for Writing Program Design at HSIs. *Composition Studies*. 45 (2): 212-215.

Falconer, Heather M. 2022. *Masking Inequality with Good Intentions: Systemic Bias, Counterspaces, and Discourse Acquisition in STEM Education*. The WAC Clearinghouse; University Press of Colorado.

Freire, Paolo. 1970. *Pedagogy of the Oppressed*. New York: Herder & Herder.

Gee, James Paul. 1989. "Literacy, Discourse, and Linguistics: Introduction – What Is Literacy?" *Journal of Education*. 171 (1): 5-25.

Gries, Laurie and Collin Gifford Brooke (Eds.). 2018). *Circulation, Writing, and Rhetoric*. University Press of Colorado.

Gonzalez, Jennifer. 2015. "Meet the Single Point Rubric." Cult of Pedagogy, Retrieved October 22, 2022. https://www.cultofpedagogy.com/single-point-rubric/.

Gonzales, Laura. 2019. "Teaching Technical Communication on the México/US Border: A Brief Case Study." Pp. 157 – 174 in *Bordered Writers: Latinx Identities and Literacy Practices at Hispanic-Serving Institutions*, edited by Isabel Baca, Yndalecio Isaac Hinojosa and Susan Wolff Murphy. New York: Suny Press.

Gopen, George and Judith Swan. 1990. "The science of scientific writing." *American Scientist*, 78 (6): 550-558.

Gusa, Diane Lynn. 2010. White Institutional Presence: The Impact of Whiteness on Campus Climate. *Harvard Educational Review*. 80 (4): 464-490.

Heath, Shirley Bryce. 1983. *Ways with Words: Language, Life and Work in Communities and Classrooms*. Boston: Cambridge University Press.

Inoue, Asao B. 2016. "Friday Plenary Address: Racism in Writing Programs and the CWPA." *Writing Program Administration*. 40 (1): 134-155.

Inoue, Asao. B. 2015. *Antiracist Writing Assessment Ecologies: Teaching and Assessing Writing for a Socially Just Future*. WAC Clearinghouse: University Press of Colorado.

Labov, William. 1972. *Language in the Inner City: Studies in the Black English Vernacular*. University of Pennsylvania Press.

Ladson-Billings, Gloria. 1995. "Toward a Theory of Culturally Relevant Pedagogy." *American Educational Research Journal*. 32 (3): 465-491.

Lu, Min-Zhan and Bruce Horner. 2013. "Translingual Literacy and Matters of Agency." Pp. 26—38 in *Literacy as Translingual Practice: Between Communities and Classrooms*, edited by Suresh Canagarjah. Routledge.

Moll, Luis, Cathy Amanti, Deborah Neff and Norma Gonzalez. 1992. "Funds of Knowledge for Teaching: Using a Qualitative Approach to Connect Homes and Classrooms." *Theory into Practice*. 31 (2): 132-141.

Paris, Django. 2012. Culturally Sustaining Pedagogy: A Needed Change in Stance, Terminology, and Practice. *Educational Researcher*. 41 (3): 93-97.

Prior, Paul. 2006. "A Sociocultural Theory of Writing." Pp. 54-66 in *Handbook of writing research*, edited by Charles A. MacArther, Steve Graham, and Jill Fitzgerald. New York: The Guilford Press.

Puzio, Kelly, Sarah Newcomer, Kristin Pratt, Kate McNeely, Michelle Jacobs, and Samantha Hooker. 2017. "Creative Failures in Culturally Sustaining Pedagogy. *Language Arts*. 94 (4): 223-233.

Russell, David. R., Mary Lea, Jan Parker, Brian Street, and Tiane Donahue. 2009. "Exploring Notions of Genre in 'Academic Literacies' and 'Writing Across the Curriculum': Approaches Across

Countries and Contexts." Pp. 395-423 in *Genre in a Changing World*, edited by Charles Bazerman, Adair Bonini and Débora Figueriredo. The WAC Clearinghouse: University Press of Colorado.

Shelton, Cecilia. 2020. Shifting Out of Neutral: Centering Difference, Bias, and Social Justice in a Business Writing Course. T*echnical Communication Quarterly*. 29 (1): 18-32.

Trimbur, John. 2000. "Composition and the Circulation of Writing." College Composition and Communication. 25 (2): 188-219.

Heikkinen, Maya. 2021. "Ungrading: Professors Experiment with Unorthodox Evaluation." *The Middlebury Campus*, March 11. Retrieved October 27, 2022. https://www.middleburycampus.com/article/ungrading-professors-experiment-with-unorthodox-evaluation.

Narrative Medicine: An Interdisciplinary Approach to Address Burnout Among the Nursing Workforce

Kimberly D. Perris < kb62@humboldt.edu > *(Cal Poly Humboldt)*, **Eden J. Donahue** *(Cal Poly Humboldt)*, **Adrian Matt Zytkoskee** *(The American University of Sharjah)*, **Janelle Adsit** *(Cal Poly Humboldt)*

Abstract

Healthcare providers (HCPs) experience unprecedented burnout. During the COVID-19 pandemic, the healthcare workforce was pushed beyond its capacity, driving some out of the field, leaving hospitals and healthcare agencies to face unrelenting demand for care. Limited staff and resources challenged organizations to redesign infrastructure and processes to meet COVID-19 safety guidelines while balancing the priorities of finance and people.

Two years into the pandemic, the signs of burnout among nurses in an RN-BSN program surfaced, which paralleled the bitter resentment happening across the nursing profession. Nurses working on the front lines reported feelings of abandonment, lack of resources, staffing shortages, exhaustion, fatigue, hopelessness, and a sense that healthcare systems were falling short in caring for the caregivers. Similar to military service members who returned from combat, nurses project the workforce will experience considerable post-traumatic stress disorder after the pandemic.

Keywords: narrative medicine, reflective practice, nursing, burnout

Reflective writing and narrative medicine provide HCPs an opportunity to listen to a patient's story profoundly and to reflect on their personal feelings and experiences. Nurses and other HCPs find validation, hope, and healing through reflection and storytelling. In partnership with narrative medicine experts in the disciplines of Psychology and English, we hope to open the door to collaboration and storytelling among disciplines and shift the culture of health care to embrace vulnerability and openness as strengths in delivering patient care.

During the COVID-19 pandemic, the healthcare workforce was pushed beyond its capacity, driving some out of the field, leaving hospitals and healthcare agencies to face unrelenting demand for care. Limited staff and resources challenged organizations to redesign infrastructure and processes to meet COVID-19 safety guidelines while balancing the priorities of finances and people. The crisis in healthcare is especially pronounced in rural areas. All who touch the healthcare system—from patients to healthcare providers (HCPs)—are affected.

Cal Poly Humboldt's nursing program is a unique site where some of the most vexing issues facing healthcare today are the focus of classroom discussions and educational practice. Cal Poly Humboldt Nursing's RN-BSN program is designed for active working nurses to achieve a Bachelor of Science in nursing (BSN). Nurses come to the program with lived experience working in the field, many seeking new opportunities and a chance to "breathe life into their career." The university nursing program has long been a fixture of our North Coast region and a site for grappling with broader healthcare needs.

For many years, Cal Poly Humboldt had a pre-licensure nursing program, where students could become a registered nurse and get a Bachelor of Science degree in nursing. In 2011, the pre-licensure nursing program was closed. Since that closure, the surrounding community and campus partners had been strategizing to create a nursing pathway at Cal Poly Humboldt. After years of planning, the university welcomed its first cohort of RN to BSN students in August of 2020. Although the world was experiencing a pandemic and the healthcare system was experiencing unprecedented amounts of stress, nurses continued to be resilient advocates for change for their patients and their profession.

As an integral part of the community, Cal Poly Humboldt has a renewed vision and purpose to address disparities across the healthcare system. As an Hispanic-Serving Institution (HSI) located on unceded Wiyot land and in close proximity to several Tribal communities including Hupa, Karuk, Mattole, Tolowa, Wailaki, and Yurok, it is the duty of Cal Poly Humboldt to continue to support students who want to serve their communities by addressing health disparities and healthcare needs. We serve our students and communities best when we prioritize the practices that honor their full humanity and their cultural diversity.

In order to address the mounting complexities of healthcare in a holistic and sustainable way, we propose reflective practice, informed by the principles of narrative medicine, as a response. We argue that reflective practice provides many benefits to the healthcare system; it not only improves the patient's experience and healthcare outcomes, but also the wellbeing of the HCP. Not only is it ethical, humane, and wellbeing-oriented to engage in this reflective work, it teaches HCPs how to practice vulnerability and communicate with patients. In essence, it is a pragmatic approach for improving healthcare outcomes, increasing efficiency, and supporting HCPs' own wellbeing and longevity in their careers. This meets the need for a more comprehensive education for HCPs—a biopsychosocial education is incomplete without reflective practice.

In what follows we identify reflective practice as an integral part of nursing education. We begin by naming the central challenges that nurses, and healthcare practitioners more broadly, face and how these have been exacerbated in a time of global crisis. We articulate the need for systemic change within healthcare and locate the role reflective practice can play in realizing widespread transformation. In particular, reflective practice may be a key means of preventing healthcare worker burnout. Drawing upon interdisciplinary scholarship, we offer a theory of reflective practice that is informed by principles of narrative medicine. We then further elaborate how the programs at Cal Poly Humboldt apply these principles in pedagogical settings—in the nursing program and in other programs within the School of Applied Health. We see the work of writing this article as itself an opportunity to enact reflective practice, as we articulate the principles that guide the nursing program's response to community and professional needs. We see this work as grounded within the Scholarship of Teaching and Learning (SoTL) as this is a self-study and an opportunity to be in conversation about the teaching and learning practices we foster in the nursing program.

Compounding Issues within the Healthcare Setting

The COVID-19 pandemic highlighted the complex issues present within the healthcare systems and nursing profession. Although this paper will focus on HCPs in a larger context, it is important to understand the dynamic of nursing as it has been and continues to be the largest section of the healthcare professions (American Association of Colleges of Nursing 2019). Many of these issues have been present since the inception of the nursing profession but are rarely discussed or researched. The image of the nursing profession has always been positive, focusing on the caring and trusting nature of nurses. For decades, society has ranked nursing as one of the most trusted and ethical professions (Saad 2022). However, despite these positive attributes, there have been increasingly challenging issues for nurses. Issues of increased work demands, lack of power/support, increased moral distress within and outside of the healthcare system, and a pervasive nursing shortage have been common themes within the nursing profession even without the additional stress of a global pandemic (Davidson et al. 2020; Haddad, Annamaraju, and Toney-Butler 2022). Nurses provide the best care possible, sometimes at the expense of their physical and mental health.

Within a healing profession, there is ubiquitous pressure on HCPs to be the epitome of health and wellness. At the same time, HCPs have often been asked to disregard their own psychosocial identities and needs (e.g., emotional, cultural, spiritual, economical, psychological) in order to more effectively address the biomedical needs of the patient. As medical doctors and faculty members of Columbia's Narrative Medicine program, DasGupta and Hurst posit that when HCPs openly grapple with personal illness (whether physical or psychological), it is often seen as threatening "the integrity of the club" and "is tantamount to treachery" to the "fraternity of healers" (DasGupta and Hurst 2007:3). Such are the pressures placed upon HCPs who deal with high levels of stress and grief, but with an imperative to not allow these forms of stress to become evident. The pressures on HCPs—and the imperatives to perform in ways that deny one's own body and mental health needs—are untenable.

As the human experience ebbs and flows, a HCP's mental health can also ebb and flow. In a workplace environment where stress and grief are regularly present, HCPs are faced with an impossible task of keeping their composure. During their shift, HCPs maintain their composure so that they are able to take care of the next patient, but what happens when HCPs go home? How does one cope with not wanting to take work home when home is the only place where one has adequate space to process the day? How does one share their experience while maintaining confidentiality? These are all questions that HCPs face. The task of processing the human experience often becomes overwhelming and can lead to physical and mental manifestations.

Healthcare providers are exposed to repeated trauma, yet are not provided with the tools and space to process emotionally challenging events. As health psychologist Nicola Davies (2021:n.p.) writes, "Many nurses agree that nurse training doesn't adequately equip them with the skills needed to cope with grief" and that "there is a discernible lack of safe spaces in healthcare institutions where nurses feel able to examine their emotions." During reflective group work within nursing populations, we have heard similar reports. One practicing nurse in the Cal Poly Humboldt nursing program, for example, reported that during the transition at the end of a shift, all the biomedical information is passed along to the next team, but no time is allotted to discuss the emotional aspects of the day. This, they noted, is particularly challenging when a patient has been abusive or passed away: there is no opportunity to process the event. The lack of psychosocial support can carry serious ramifications for HCPs.

One such ramification is addiction. Without healthy outlets for addressing the challenging emotions associated with regular exposure to trauma, it is easy for HCPs to turn toward numbing agents in order to self-medicate and cope. For instance, in working with emergency medical technicians (EMT) responders in reflective writing groups, it is common to hear reports of going out to drink after every shift. To illustrate, after sharing with the group a story involving the death of a two-year old boy, an EMT participant shared, "I think everyone wanted to cry, but crying isn't okay, so we got hammered after work and made horrible jokes in some kind of attempt to not feel it." The challenge, he noted, was knowing this was not healthy but feeling unable to cope with it differently because "it's just how it works in this profession." Speaking to this, Monroe and Kenaga (2011:1) note that instances of addiction within the nursing field are estimated to be as high as 20%, yet a culture of silence and fear of punishment often prevent nurses from seeking help. Furthermore, the easy access to prescription medication in healthcare contexts only exacerbates the risk. While substances may provide temporary relief, they are no substitute for processing challenging experiences.

When HCPs experience mental health issues themselves, they often turn inward due to fear of stigmatization; yet, ironically, if a patient had mental health issues, the HCP would help that person navigate to support and resources. According to HCP researchers Davidson et al. (2020), the U.S. had already seen increasing rates of suicide among HCPs before the pandemic. They recognized the toll of these stressors and conducted the first longitudinal study to examine the complexities of nurse suicide in the U.S. from 2005 to 2016. They found that compared to the general population, nurses were not only at a greater risk of suicide, as previous researchers had shown, but nurses completed suicide at a higher rate than the general population. Comparatively, nurses were also more likely to have a mental health history and issues within the workplace (Davidson et al. 2020). In addition, nurses might delay mental health care due to workplace and professional stigma (Knaak, Mantler, and Szeto 2017; Gilligan 2021). The stigma is perpetuated by workplace and professional cultures that portray HCPs who need mental health care as people who are unable to work in a high stress environment (Knaak et al. 2017). Given the immense amount of stress that HCPs experience, there is a reluctance to discuss mental health concerns, which can lead to isolation, reliance on other coping mechanisms (e.g. self-medicating), and burnout (Knaak et al. 2017).

Two years into the pandemic, the signs of burnout among nurses have surfaced with increasing frequency and bitter resentment. Nurses working on the front lines reported feelings of abandonment, lack of resources, staffing shortages, exhaustion, fatigue, hopelessness, and a sense that healthcare systems were falling short in caring for the caregivers. The increased stress and unaddressed mental healthcare crises from the COVID-19 are projected to exacerbate mental health issues and suicide rates among healthcare workers (Awan et al. 2022).

The influence of the workplace environment on a HCP's experience, both as a provider and as an employee, cannot be overlooked, especially as retention of the healthcare workforce is even more imperative in times of ongoing public health crisis. In a systematic review, Yahyaei et al. (2022) examined factors within the workplace environment and nurses' intention to stay within that organization. One of the main elements that increased nurses' intention to stay was organizational support and nursing empowerment (Yahyaei et al. 2022). The importance of nursing empowerment has been a key component of the nursing profession both historically and currently (Kagan 2006). JoAnn Ashley, a nursing scholar and advocate, defined power in nursing as being composed of awareness, freedom, choice, action, and creativity (Kagan 2006). Ashley was also aware of the complex relationships between the nursing profession, other healthcare professions, society, and the healthcare system. They are separate but overlapping entities that must allow for intrinsic growth separately and together, all while grounded in humanistic principles (Kagan 2006). Yahyaei et al. (2022) focused on nurses' intention to stay within an organization, rather than the profession. Yet in rural areas with a lack of nurses and healthcare organizations, nurses have limited options when their workplace environments are not supportive. Organizations have limited access to qualified applicants when nurses leave the organization or the profession. From bedside nurses to administrators, the cyclical turnover of nurses negatively impacts the entire healthcare system.

Education Toward a Culture of Caring in Institutions of Health

Systemic change in healthcare settings may start with the reflective practice of the individual. Nurses report nursing school does not adequately prepare them for the emotional demands, complexities of patient care, and human factors they face in practice. As such, most nursing programs do not integrate self-care strategies to manage compassion fatigue, the demands of working in complex settings, and working with multiple providers and disciplines (Horton-Deutsch and Sherwood 2017). Transforming healthcare systems requires a shift in education that includes innovative practice that fosters human caring. Jean Watson's theory of human caring offers a framework for teaching and learning as a caring encounter, "embracing the belief that trust and faith in human expression and self-actualization is the focus of the educational process" (Hills and Watson 2011:218). The premise of Watson's caring science model is caring for self to care for others. Reflective practice shifts the paradigm of traditional nursing education from a lecturing model to a learner centered education that honors and leverages the experience of the individual to foster professional growth and development (Horton-Deutsch and Sherwood 2017). Through transformational practice nurses reflect and apply learning, thus developing multiple ways of knowing beyond empirical knowledge that includes aesthetic, legal, ethical, and personal knowledge (Horton-Deutsch and Sherwood 2017). The scholarship of teaching and learning within healthcare disciplines can continue to realize the promise of these educational shifts, as healthcare pedagogy continues to respond to the pressing needs of the practitioners we serve in the health disciplines.

Healthcare Worker Burnout

Compassion is core to nursing practice and is defined by the American Nurses Association Code of Ethics as "An awareness of human suffering, tempered with reason, coupled with a desire to relieve the suffering; a virtue combining empathy, benevolence, caring, and mercy. Used with the cognitive and psychomotor skills of healing to meet the patient's needs" (ANA 2015:41). Unfortunately, when nurses face continuous trauma without adequate support, they may be exposed to secondary traumatic stress resulting in feelings of despair that contribute to physical symptoms including anxiety, sleeplessness, lack of job satisfaction, nausea, and depression (Schmidt and Haglund 2017). Ultimately, without intervention, these symptoms can lead to burnout (sometimes referred to as "compassion fatigue"). Burnout, as physicians Gregory Nolan and colleagues note, can be "characterized by depersonalization, emotional exhaustion, and low personal accomplishment" and result in "decreased productivity, depression and suicidal ideation, impaired professionalism and communication, and professional errors and near misses" (2020:184). We have also heard HCPs describe burnout as apathy, disillusionment, and even anger with patients.

Traditionally, it has been suggested that the best method for avoiding burnout and establishing professionalism is to maintain emotional detachment from both patients and self. Yet, research increasingly supports that the practice of "self-reflection and the ability to constructively process emotions" serve to protect against stress and burnout (Guillemin and Gillam 2015:727). Put simply, the practice of detaching from emotional engagement is not the safeguard it once was thought to be. In fact, it can serve to generate a range of serious threats to HCPs' wellbeing. For instance, when HCPs' emotional lives and reflections on their workplace experiences are seen as irrelevant or even inappropriate to the work they perform, it can create social alienation and disillusionment. Kinman et al. (2016:8) describe these types of psychosocial ramifications involved in nursing labor and call for "social support" in which there are "opportunities for staff to create space for reflection, discuss emotional reactions to practice, and reaffirm core nursing values." Doing so, they argue, will help "protect against social alienation and burnout" through the development of "social connectedness" (2016:8).

Put simply, the cause of HCP burnout is a result of multifaceted systemic dysfunction causing a cascade of negative outcomes on the overall health of the population and the wellbeing of those who care for them. The U.S. Surgeon General, who names lack of a space for vulnerability as a factor contributing to burnout, has established addressing HCP burnout as a national priority (U.S. Department of Health and Human Services 2022). These issues within the healthcare system are intersectional. Certain positions within the healthcare sector employ higher numbers of women and people of color—people who find the challenges of healthcare service compounded by racism, classism, sexism, and a historical lack of power. These issues are not only present within the healthcare system and society but within healthcare organizations and workplace cultures. There is an urgent need to create work environments that foster a culture of collaboration and vulnerability and allow for individuals to have their voice heard.

Reflective Practice as a Response

As a response to these difficult circumstances facing HCPs, including those of the students who come to Cal Poly Humboldt's newly formed School of Applied Health, we seek to equip students with the skills of resilience, self-acknowledgement, and self-care that are cultivated in reflective practice. We take this orientation with the research-informed perspective that reflection is both good for HCPs and good for their patients. Reflective practice is therefore the center of our pedagogy.

Boud, Keogh, and Walker (1999) define reflective practice as "those intellectual and affective activities in which individuals engage to explore their experiences in order to lead to a new understanding and appreciation" Drawing from their definition, by "reflective practice" we mean an orientation and way-of-being that guides one's decision-making and supports one's resilience in each encounter. In other words, reflective practice is more than a simple skill-set; instead, it might be thought of as a deliberate disposition or habit-of-mind that translates into one's day-to-day communications, attitudes, and choices.

To elaborate, reflection provides the metacognitive possibility of making one's interiorized thought-life less automatic and more available for consideration and, in turn, revision. Reflection allows for re-evaluation and the potential to change one's patterns. For these reasons, reflective practice is associated with a growth mindset that orients toward learning—finding the opportunities available in challenging and changing circumstances. Speaking to this, Lutz et al. (2013:343) found that their reflective practice pedagogy enabled students to "come to see difficult professional situations as challenging yet manageable learning experiences, thus leading to the personal outcome of a sense of improved attitudes and skills for the management of difficult personal and interpersonal situations and reduced stress."

Reflective practice is thought to cultivate resilience; challenges are met with flexibility, responsiveness, and an ability to adapt. This orientation also requires a degree of detachment; difficult circumstances are expected and met with confidence, and difficulty is not judged as "bad" or as a reflection on the self. A reflective practice refuses self-judgment or fixed labeling, but instead seeks to better attune to the situation as-is, in order to serve the needs (of self and other) in a given circumstance. In this sense, reflective practice is related to mindfulness—the cultivation of nonjudgmental awareness—which has been shown to have positive effects for wellbeing. Research on the effectiveness of mindfulness based stress reduction (MBSR) continues to proliferate within the healthcare sector (van der Riet, Levett-Jones, and Aquino-Russell 2018; Gautam, Palaniveu, and Kaur 2020; ElKayal and Metwaly 2022; Wexler and Schellinger 2022). It is for these reasons that we turn to reflective practice as a way-of-being, one that can support the HCP who navigates the adverse circumstances of today's healthcare system.

Reflection is a deliberate practice and continually cultivated skill, rather than an assumed human quality. In a longitudinal study of 117 medical students, Park et al. (2022: 1) found that "[s]elf-reflective ability is not naturally developed as students' progress through grade levels." They recommend: "Educational intervention is needed to help students understand approaches to self-reflection and its importance in enabling them to develop their abilities as well as to participate actively in reflective writing" (1).

There are many pedagogies of reflection utilized in health-related professional training. For example, Lutz et al. (2018) used clinical reflection training (CRT) which focuses on experiential learning centered on the professional dilemmas faced by medical students. The features of a CRT pedagogy are a focus on current and real problems; an experienced and supportive trainer; a supportive group with whom reflection could be processed in dialogue; and a secure space for reflection. The pedagogy invites students to observe their bodies, their emotions, and their mental models with curiosity and non-attachment. Lutz et al. (2018) report that participants experienced stress reduction and reported improved quality of patient care as a result of the reflective practice pedagogy.

Reflective practice informed by the principles of narrative medicine

Within medical education, reflective practice has been cultivated through the research and application of narrative medicine or narrative-based medicine. Narrative medicine is a field that cultivates a set of skills and experiences for both patients and HCPs, seeking to make healthcare more humane, empowering, and beneficial for all.

As Rita Charon (2017:1) explains, "Narrative medicine began as a rigorous intellectual and clinical discipline to fortify healthcare with the capacity to skillfully receive the accounts persons give of themselves—to recognize, absorb, interpret, and be moved into action by the stories of others." Charon, an MD and PhD in literature, is most often attributed as founding and solidifying the field of narrative medicine in the mid-nineties. She writes of the emergence of the field, from her seminal work *Narrative Medicine: Honoring the Stories of Illness* (2008: vii): "The field of narrative medicine has emerged gradually from a confluence of sources—humanities and medicine, primary care medicine, contemporary narratology, and the study of effective doctor-patient relationships." She argues that the value of narrative medicine is that it "provides health care professionals with practical wisdom in comprehending what patients endure in illness and what they themselves undergo in the care of the sick." The field has burgeoned to include many sub-areas, including practices that enable better patient-provider relationships, more ethical orientations, career longevity and wellbeing for healthcare workers.

Although its influence continues to grow, narrative medicine is primarily associated with the medical humanities, a disciplinary home for multiple forms of scholarship and practice that explore the ethical, historical, literary, creative, anthropological, philosophical, and religious dimensions of medicine or health. The medical humanities are guided by the recognition that medical practice always involves interpersonal communication, interpretation, ethics, empathic orientations, and humanistic understanding.

Narrative medicine is, in Sayantani DasGupta (2022:n.p.) words, "the clinical and scholarly endeavor to honor the role of story in the healing relationship." The field understands story as permeating every part of the clinical encounter—from how a patient tells of their symptoms to how HCPs tell the story of treatment and prognosis.

Story can also be a deliberately utilized tool for stress reduction, reflection, and coping for both patients and HCPs. Reflective writing has an important role in narrative medicine and opportunities to engage in it have proliferated throughout a broad range of healthcare contexts, including hospitals and higher education.

In what follows, we demonstrate how reflective practice, informed by the principles of narrative medicine, supports the HCP's ability to attune to the stories occurring in the clinical setting, enabling providers to become better able to treat patients with accuracy and care. At the same time, reflective practice also supports HCPs' own wellbeing. Reflective writing is one avenue for engaging with a reflective practice, and it can support self-understanding, acceptance, growth, and resilience in the face of the challenging circumstances every HCP encounters.

Reflective practice and narrative medicine in the clinical setting: toward patient-centered care

Medical practice is always mediated by interpretation and story. The patient tells the story of their symptoms, their background, their cultural orientation to treatment; this story is interpreted and acted upon by the HCP. In turn, the HCP tells a story of diagnosis—which is a story that imagines the past, present, and future of a disease or condition. As Lewis Mehl-Madrona (2007:83) notes, in clinical settings, we are always co-creating the story: "We are co-creating a shared story of healers and patients/families/communities wherever we go. We are immersed in the art of storytelling." It is imperative that the HCP be able to listen to the patient's story and to reflect on their interpretation of the story, questioning assumptions and seeking greater understanding.

This type of reflection requires humility, a point emphasized by Melanie Tervalon and Jann Murray-García (1998) in their scholarship and training on cultural humility. Noting the "sociocultural mismatches" that occur between patients and HCPs, along with the systemic prevalence and internalized biases of racism, classism, homophobia, and sexism, Tervalon and Murray-García (1998) called for an intervention and reorientation in how HCPs approach clinical communications. Their work undoes medical paternalism as Tervalon and Murray-García call upon HCPs to commit to patient-centered care that prioritizes the agency, cultural orientations, and needs of the people whom healthcare is meant to serve. Cultural humility is a reflective practice as it requires, in Tervalon and Murray-Garcías' words (1998:123), "a lifelong commitment to self-evaluation and critique, to redressing the power imbalances in the patient dynamic, and to developing mutually beneficial and non-paternalistic partnerships with communities on behalf of individuals and defined populations".

Building upon Tervalon and Murray-García's work, a specifically *narrative* humility has also been theorized by Sayantani DasGupta, drawing upon her background in narrative medicine. With the term "narrative humility," DasGupta emphasizes several forms of recognition that HCPs should practice in the clinical encounter. Specifically, the recognition that our patients' stories and their health conditions "are not objects that we can comprehend or master, but rather dynamic entities that we can approach and engage with"; respect for complexity, ambiguity, contingency, and contradiction as always present in the human condition; readiness to creatively co-create knowledge and meaning in the clinical space through acknowledgment of multiple perspectives and the authorization of the patient's knowledge and decision-making; and prioritization of self-evaluation and self-critique about our roles, expectations, responsibilities, attachments, identifications, and limits in any given situation. Narrative humility includes the moment-by-moment self-questioning: What/who am I valuing or prioritizing? What am I assuming or believing to be true?

A reflective and humble orientation counters several assumptions within medical practice: 1) the false assumption that medicine is only about finding the answers to scientific questions and "solving" healthcare conditions; it is not about people's identities, emotions, cultures, and relationships; 2) the false assumption that the treatment plan should be developed solely by the doctor without patient input and consultation; and 3) the false assumption that the provider is the sole expert and knows best on behalf of the other. In other words, narrative humility seeks to counter the institutionalized medical paternalism that has done much harm within the medical system.

Because of its demonstrated outcomes, reflective practice, as the cornerstone of narrative medicine, is a skill that is promoted across many medical education programs. Daryazadeh et al. (2020) conducted a quasi-experimental study of 135 medical interns and found that those who received narrative-based education increased their reflective capacities beyond those of the control group. With the addition of seven two-hour reflective practice sessions, a significant difference emerged in participants' scores on the Reflection Evaluation for Learners' Enhanced Competencies Tool (REFLECT) rubric (Daryazadeh et al. 2020).

George Zaharias (2018) notes that narrative practice "is intrinsically therapeutic for the patient (in the telling and in being listened to)." The skillset of close listening cultivated in the literary classroom, for instance, translates to the close listening needed to recognize that, in the words of Melanie Gregg (2020:n.p.), "Patients are not their symptoms... They are whole human beings with entire life experiences. Having the ability to listen with empathy and interpret and pull meaning from it is how a caregiver can lead a patient out of suffering and into a path of healing."

Healthcare providers who take a narrative medicine approach may use a number of techniques to enable "close reading" and "close listening" of patients' stories. One such technique is the composing and sharing of "parallel charts." A parallel chart is a log of writing reflections that give a window into the psychosocial thoughts and feelings arising during patient interactions. While such records are not conventionally found in medical charts, they nonetheless can inform skillful care. The patient may also be supported by sharing selections of this writing within a small group of peers (Charon 2006:156). Parallel charts may take any number of forms; the emphasis is not on a particular length or form, but instead on the regularity of the writing, as a tool for supporting the ongoing communication between patient and HCP and the opportunity for self-reflection from both participants in the medical relationship.

To provide an illustration, Matt (co-author of the present text) worked with a young social work student who had an internship at an addiction recovery center. Though she liked the work, she found herself particularly uncomfortable with one of the residents. It was not until she sat down and wrote her parallel chart that she realized the man reminded her of one of her mother's abusive boyfriends from her childhood. In this case, her parallel chart revealed an inherent bias previously not recognized. As a result, her reflection alleviated some guilt she had been experiencing for her feelings and, as a result, allowed her to process her interactions with the man more mindfully. Often cited as the most meaningful part of the process, part two of her parallel chart process was to read it aloud to a group of peers. It is crucial to note that she was not required to do so ("passing" is always an option) and the feedback she received from other group members was limited to supportive expressions of gratitude and/or questions grounded in curiosity. This is *not* a space for giving advice, comfort, or critique. It is, rather, a space to listen and be heard and to bear witness to each other's experiences. It is a space to grow personally and professionally through reflection.

The "illness narrative" is another genre that promotes reflective practice through writing—one that has been used by narrative medicine practitioners Sayantani DasGupta and Rita Charon (2006) of Columbia University's Medical Center. Common goals of this practice are to: 1) situate participants in a patient's perspective through the practice of vulnerability; 2) develop trust in teammates while practicing compassionate communication; and 3) explore one's own relationship with illness and health in order to move against the biomedical tradition of silencing healthcare provider's biopsychosocial realities.

Matt has engaged illness narratives in medical education settings (Zytkoskee 2020:178), using this genre to support HCPs in reflecting on their own identity within the field. By gaining a better sense of who they are as professionals, why they entered the field, and why they are thinking, feeling, and behaving in certain ways, they are more likely to thrive. Illness narratives give providers an opportunity to reflect on the times when they have been in the role of patient within a healthcare setting, or when they have been the witness to another patient's experience from a position of family member or loved one, rather than the HCP. Such reflection cultivates empathy for the patient as it also fosters self-understanding.

Matt's experience working with HCPs included a multi-step process that models a common way of scaffolding provider engagement with narrative medicine approaches. First, participants come to an orientation session in which they ask questions and discuss example illness narratives that are provided ahead of time. Then they begin the writing process through brainstorming activities such as freewriting (writing non-stop for short periods of time), listing, and sensory description (reflecting on an experience through the lens of all five senses). Following this introductory session, participants independently write their illness narratives, which are defined as accounts of illness/injury they have experienced directly or by someone with whom they had a personal relationship. The primary guideline is that narratives must be about how the experience affected the participant personally. Once the narratives are completed, participants meet in small groups of four to ten and read their narratives aloud. With any group-sharing process within a narrative medicine framework, feedback should be limited to supportive comments and/or questions grounded in curiosity (e.g."How old were you when this happened?"). It is important to remember that, while the texts produced are likely to be rhetorically powerful, the setting is not a creative writing class focused on an end product. Rather, it is about the benefits the act of writing has for the writer (Yagelski 2009:9). When reflective writers are given the opportunity to connect and share their writing with peers in meaningful, safe contexts, it can help them "develop the capacity for empathetic listening" and gain "significant insight" into "shared humanity" (Sierpina 2007:628). In essence, both the act of writing and the sharing of that writing carry significant potential for individual and collective growth.

Reflective practice and The Center for Mind-Body Medicine training in the clinical setting: toward HCP wellbeing

Another framework for the development of reflective practices is found in the Center for Mind-Body Medicine's (CMBM) evidence-based curriculum. CMBM was founded by Georgetown Medical School psychiatrist and former Chairman of the White House Commission on Complementary and Alternative Medicine Policy, James S. Gordon (2009). It focuses on self-care, social support, and community building in a broad range of populations and ages. Mind-Body Skills Groups, which espouse the mission of "teaching thousands to heal millions" (CMBM 2023), focus on teaching participants practices grounded in meditation, reflective writing, drawing, guided imagery, autogenics, biofeedback therapy, mindful eating, mindful movement, and genograms (category-focused family trees). Research supports the efficacy of these groups (Maclaughlin et al. 2011; Greeson, Toohey, and Pearce 2015). Healthcare professionals experience significant declines in emotional exhaustion, depersonalization, and secondary traumatic stress, and improvements in sense of personal accomplishment and satisfaction with work (Weinlander, Gaza, and Winget 2020). A particularly appealing aspect of their approach to self-care and wellbeing is found in its straight-forward and pragmatic nature—the skills, once learned within a group setting, are easily practiced alone. To briefly illustrate, we will now outline two practices taught within CMBM groups.

To begin, CMBM offers a skill they term, "The Wise Guide," a guided imagery practice designed to help the HCP establish a safe psychological space for processing human experiences and accessing inner knowledge (Gordon 2009:117). Typically, this technique is facilitated within a mind-body skills group but, as noted before, may be practiced independently once an understanding of the process is rooted. It begins with a concentrative meditation technique, "soft belly breathing," which activates the parasympathetic nervous system (Gordon 2009:44). Next, participants are guided through a visualization of an internal "safe place," a place where they may find peace and reassurance even when facing trauma—whether past or present (Gordon 2009:114). Once the participant is settled in this internal safe zone, they are invited to "meet their wise guide." What this wise guide represents depends entirely upon the individual, with concepts ranging from a spiritual manifestation, a higher power, a collective consciousness, the imagined wisdom of someone in one's life, or a manifestation of one's own unconscious wisdom. Once the wise guide arrives, it is time to ask it questions regarding life circumstances, experiences, and trajectories. If responses are confusing, participants ask for clarification and accept that "words, images, sounds, feelings, impressions" all potentially hold answers (Gordon 2009:119). This dialogue is where processing takes place, with many participants reporting that they received answers and insights they had not expected. The process lasts for roughly twenty minutes. After participants are gently guided back to group consciousness, they are given twenty minutes to write reflectively and/or share their experiences as a way to retain and process.

Another reflective practice taught within CMBM's Mind-Body Skills groups is the use of drawing as an outlet for self-expression, self-reflection, and an avenue for accessing the subconscious in order to find self-awareness and healing. LaPenna and Tariman note that art therapy is "an integrative mental health intervention involving creative processes, application of psychological theories, and human experiences within a psychotherapeutic relationship," and has an extensive history in the treatment of trauma with results demonstrating improvement in patient anxiety, depression, and overall quality of life (Gordon 2009). Supported by research on the efficacy of art therapy to improve both psychological and physical trauma (Kaimal, Ray, and Muniz 2016; Hass-Cohen et al. 2019; Betensky 1995; Broadbent et al. 2018), CMBM's approach employs drawing in the first and last sessions of their eight-week curriculum. In session one, they give participants roughly five minutes per drawing to complete three drawings with brief directions between each one encouraging participants to let the drawings manifest uncensored and quickly in order to be "authentic, surprising, revealing" (Gordon 2019:61). The prompts, which are designed to guide exploration of one's psychosocial realities, obstacles, and trajectories include: 1) draw yourself as you are now; 2) draw yourself with your biggest problem, challenge or issue; and 3) draw yourself with your biggest problem, challenge or issue solved (Gordon 2019:62-63). Once all three drawings are complete, individuals briefly present each drawing to the group and describe any sights, thoughts, and/or emotions related to the drawings and the experience of creating them. Facilitators are trained to guide this process through questions intended to help participants explore the experience at a deeper level (if needed). Analysis and critique are not part of the process other than by the participants themselves. Ultimately, the goal is that participants emerge from this reflective experience with a greater understanding of their psychosocial realities and how they would like to guide their lives moving forward.

Reflection as a Way to Humanize Nursing Educaion at Cal Poly Humboldt

The current Cal Poly Humboldt nursing program launched in 2020 with a cohort of thirteen students. The courses are taught mostly online to support flexibility for working nurses. Students and faculty come together face-to-face twice a semester for weekend intensives. These weekends provide students with an immersive community building and learning experience. Through presentations and workshops from the faculty, community, national experts, and each other, they learn about topics related to their courses. The intensives also provide a forum to analyze the complex issues they face in serving diverse populations. These conversations encourage them to evaluate system-wide challenges and disparities and envision the healthcare system they want to create.

An innovative approach to strengthen bonds among Cal Poly Humboldt nurses

Given our rural regional context, Cal Poly Humboldt has a responsibility to address healthcare needs across multiple dimensions in a primarily online learning environment. Challenges with online learning include potential feelings of isolation among students. Cal Poly Humboldt's nursing program is intentional about creating a humanized approach to online learning. The program is mostly asynchronous, with two face-to-face weekend intensives each semester. We integrate reflective writing within discussion boards that includes student analyses and written and video reflections on the stories from their peers. The weekend intensives are also key to this goal. During year-one, the height of the pandemic, all the intensives were held synchronously online via Zoom. The bonds between students were strengthened through real-time interactions with each other. Since then, we have shifted weekend intensives to originally envisioned face-to-face experiences.

Fostering vulnerability through Narrative Medicine workshops

A program feature is designated time for writing and reflection that allows nursing students to process experiences in their workplaces. During these writing workshops, nursing students have shared how rare and important these opportunities are. One practicing nurse in the Cal Poly Humboldt nursing program, for example, reported that during the transition at the end of a shift, all the biomedical information is passed along to the next team, but no time is allotted to discuss the emotional aspects of the day. This, they noted, is particularly challenging when a patient has been abusive or passed away: there is no opportunity to process the event. The lack of psychosocial support can carry serious ramifications for HCPs.

During intensives, students attend narrative medicine workshops where they are invited to participate in reflective writing exercises. A short story exercise, for example, is designed to encapsulate key experiences of HCPs (Fogarty 2009). Students were asked to reflect on and write down a list of impactful illness/injury/trauma they had experienced. Through a series of brief free writing activities, they then identified words and emotions that were important to relating the experiences. Finally, they edited the writing down to a fifty-five word story. Students then shared their story with a peer in an online breakout room. Some students shared deeply personal experiences they had with patients and ill family members. This exercise created space for students to be vulnerable with their peers while having an expert in the field to help facilitate and debrief afterwards. Through this type reflective dialogue, both online and in person, students report feeling a sense of community in knowing they are not alone in feeling overwhelmed and stressed out by school, work, and family.

Reflective pedagogy in the nursing program at Cal Poly Humboldt

The start of 2020 began with the International Year of the Nurse to honor the 200th birthday of Florence Nightingale, the mother of nursing. This was a year to honor the physically, emotionally, and intellectually challenging work that nurses endure to care for their clients. This worldwide recognition coincided with the COVID-19 pandemic, shining a spotlight on the essential role of nurses at the cost of unrelenting physical and emotional stress. By reframing experiences through CARING values,[1] students are able to contextualize their day-to-day experiences through analysis and reflection. The design of the virtual classroom and practical learning experiences are grounded in a humanized approach to foster community, connection, and innovation.

The COVID-19 pandemic forced educators to rethink the way they teach and the way they engage their students. Disruptive innovation inspired nursing program faculty to embrace online tools and resources in designing the new program curriculum and modes of delivery. The virtual classroom and practical learning experiences fostered inquiry and empowerment relevant to each student's nursing practice. In developing the first courses, the nursing faculty promoted a culture of caring while acknowledging the experience of frontline nursing students in the pandemic. The faculty designed cohesive learning environments where nursing students reflected on their lived experiences as nurses. Students leveraged new skills to foster inquiry and empowerment in their work in real time and with future-focused goals. Our nursing courses continue to have asynchronous and synchronous sessions to allow flexibility for working nurses. In order to foster a supportive class culture in a hybrid format, nursing students create written analyses and spoken reflections of their lived experiences.

In the first semester of the nursing program, students work on a reflective paper where they think about how people experience illness and the dynamic of power (or powerlessness) of nurses within the healthcare system. Students read seminal articles by Rita Charon (2009) and Paula Kagan (2006), as well as stories from patients with a chronic illness. Students then reflect on the role of the nurse and their own expectations of patients, their families, and HCPs. Shortly after submitting this paper, the students participate in a writing session facilitated by an interdisciplinary colleague. Although the incoming cohort completes the paper, the reflective writing session is conducted with both the incoming and continuing cohorts. The continuing students are included in the reflective writing session to encourage the adoption of reflective practice within their day-to-day nursing practice. Reflective writing is also built into each of the courses and discussion boards assignments.

One such example was a discussion board assignment focused on strategies for success to prevent burnout. Students wrote about their preferred communication style and how they could address burnout on a personal and professional level. Students then recorded a spoken reflection about the art of nursing and how they would nurture self-care during the nursing program and beyond. By recognizing communication styles and burnout rates, students are able to express their experiences on a personal and professional level to grow into more holistic, nurturing providers. One of the students shared: "It's kinda like therapy for all of us to come together and talk about these stressful experiences we have as nurses." Through reflection, students are able to honor the full spectrum of emotions that do not have a predetermined path, from catharsis to doubt. Students are able to learn the language and skills to process emotions as there is often a cascading effect. The emotional journey might prompt students to ask questions about themselves and their processing. A common question for HCPs is: "if I leave my emotions at the door, does it make me a bad person?"

Another way that the nursing program is fostering reflective practice is with virtual reality. Students experience nursing simulations to practice skills without the risk of adverse outcomes to a real patient. Although nursing education has adopted various innovative practices over the years, there was always a focus on the perspective of the nurse. One way to foster empathy and reflective practice is to have people experience another's reality. In the past, this was accomplished through personal experiences with the healthcare system and/or trying to view a situation from the perspective of the patient. The adoption of virtual reality allows nurses to experience the patient perspective of health, illness, and the healthcare system. Throughout the immersive

1. CARING values: Collaboration & cultural humility, Advocacy, Rurality, Innovation & openness, Nurture - self & those we care for, Growth & life-long learning

experience, students navigate the complex emotions of being a patient. Nursing faculty then provide a guided debriefing experience so that students can discuss the various experiences with their classmates. After the virtual reality simulation, students also write about the experience to further the reflection process with such questions as: how did you feel, was it different from what you expected, and how will you grow from these experiences? Often, these types of questions or reflective practices can be incongruous with the apparatus of the education and healthcare systems, which focus on evaluation, assigning, testing, and the separation of providers from the patient experience. Students must be fully engaged with the education, for a transformative educational experience where students broaden their knowledge, skills, and attitudes/dispositions.

The content within the coursework provides students with the multifaceted experience so they can foster a culture of healing, community, and connection within their workplace and professions through reflective practice. Although the nursing program facilitates reflective practice through activities and assignments, students need to continue these activities outside of class for reflective practice to become part of their routine. Through reflection, HCPs are able to find a safe space to honor and release emotions that arise while caring for people who experience the miracles and tragedies of life.

Going forward, we want to cultivate cross disciplinary partnerships with narrative medicine experts in Psychology and English to compile and better understand stories from nurses. We hope that this study of our own students and our teaching might open the door to collaboration and storytelling among other disciplines. Ultimately, we want to shift the culture of care across applied health fields to embrace vulnerability and caring as a strength in delivering patient care.

References

American Association of Colleges of Nursing. 2019. "Nursing Fact Sheet." Washington, DC: American Association of Colleges of Nursing. Retrieved Sept. 2, 2022. https://www.aacnnursing.org/News-Information/Fact-Sheets.

American Nurses Association. 2015. *Code of Ethics for Nurses*. Silver Springs, MD: American Nurses Publishing.

Awan, Sana, Mufaddal N. Diwan, Alifiya Aamir, Zoha Allahuddin, Muhammad Irfan, Alessandro Carano, Federcia Vellante, Antonio Ventriglio, Michele Fornaro, Allesandro Valchera, Mauro Pettorruso, Giovanni Martinotti, Massimo Di Giannantonio, Irfan Ullah, and Domenico De Berardis. 2022. "Suicide in Healthcare Workers: Determinants, Challenges, and the Impact of COVID-19." *Frontiers in Psychiatry 12*. https://doi.org/10.3389%2Ffpsyt.2021.792925.

Betensky, Mala G. 1995. *What Do You See? Phenomenology of Therapeutic Art Expression*. London, England: Jessica Kingsley Publishers.

Boud, David, Rosemary Keogh, and David Walker. 2013. *Reflection: Turning Experience into Learning*. London and New York: Routledge Falmer.

Broadbent, Elizabeth, Jan W. Schoones, Jitske Tiemensma, and Ad A. Kaptein. 2018. "A Systematic Review of Patients' Drawing of Illness: Implications for Research Using the Common Sense Model." *Health Psychology Review* 13(4):1–21. https://doi.org/10.1080/17437199.2018.1558088.

California Legislative Senate Bill 628. California Creative Workforce Act of 2021. Chaptered date: 10/09/21. https://leginfo.legislature.ca.gov/faces/billStatusClient.xhtml?-bill_id=202120220SB628.

Center for Mind-Body Medicine. 2023. "About CMBM." Washington, D.C.: The Center for Mind-Body Medicine. Retrieved January 10, 2023. https://cmbm.org/.

Charon, Rita. 2008. *Narrative Medicine: Honoring the Stories of Illness*. New York: Oxford University Press.

Charon, Rita. 2009. "Narrative Medicine as Witness for the Self-Telling Body." *Journal of Applied Communication Research* 37(2):118-131. http://dx.doi.org/10.1080/00909880902792248.

Charon, Rita. 2017. "Introduction." *The Principles and Practices of Narrative Medicine*. Eds. Rita Charon, et al. New York: Oxford University Press.

Clance, Pauline R. and Suzanne A. Imes. 1978. "The Imposter Phenomenon in High Achieving Women: Dynamics and Therapeutic Intervention." *Psychotherapy: Theory, Research & Practice* 15(3):241–47.

Daryazadeh, Saeideh, Payman Adibi, Nikoo Yamani, and Roya Mollabashi. 2020. "Impact of a Narrative Medicine Program on Reflective Capacity and Empathy of Medical Students in Iran." *Journal of Educational Evaluation for Health Professions* 17(3):1-7. https://doi.org/10.3352/jeehp.2020.17.3.

DasGupta, Sayantani. 2008. "The Art of Medicine: Narrative Humility." *The Lancet* (British edition) 371(9617):980–981. https://doi.org/10.1016/S0140-6736(08)60440-7.

DasGupta, Sayantani. Personal Website. 2022. http://www.sayantanidasgupta.com/doctor/about.

DasGupta, Sayantani and Marsha Hurst. 2007. *Stories of Illness and Healing: Women Write Their Bodies*. Kent, Ohio: Kent State University Press.

DasGupta, Sayantani and Rita Charon. 2004. "Personal Illness Narratives: Using Reflective Writing to Teach Empathy." *Academic Medicine: Journal of the Association of American Medical Colleges* 79(4):351–56.

Davidson, Judy E., James Proudfood, Kelly Lee, Garni Terterian, and Sidney Zisook. 2020. "A Longitudinal Analysis of Nurse Suicide in the United States (2005-2016) with Recommendations for Action." *Worldviews on Evidence-Based Nursing* 17:6-15. https://doi.org/10.1111/wvn.12419.

Davies, Nicola. 2021. "Understanding Grief in Nursing." *Independent Nurse for Primary Care and Community Nurses*, May 13. Retrieved July 6, 2022. https://www.independentnurse.co.uk/professional-article/understanding-grief-in-nursing/237189/.

ElKayal, Manal Mohamed and Safaa Mohamed Metwaly. 2022. "Effectiveness of Mindfulness-Based Intervention on Post-Traumatic Stress Symptoms Among Emergency Nursing Students." *Middle East Current Psychiatry* (Cairo) 29(1):1–13.

Fancourt, Daisy and Saoirse Finn. 2019. "What Is the Evidence on the Role of the Arts in Improving Health and Well-being? A Scoping Review." *Health Evidence Network Synthesis Report* 67. World Health Organization.

Fogarty, Colleen T. 2009. "55-Word Stories: Introducing the 55-Word Stories Feature Section." *Families, Systems, & Health* 27(3):284-285.

Gautam, Anamika, Srinivasan Palanivelu, and Gurvinder Kaur. 2020. "Effectiveness of Mindfulness Based Stress Reduction Program (MBSR) on Stress Among Nursing Students- a Mixed Method Study." *International Journal of Community Medicine and Public Health* 7(9):3547–3552. http://dx.doi.org/10.18203/2394-6040.ijcmph20203921.

Gilligan, Heather T. 2021. "Nurses at High Risk for Suicide: 'I just wanted all of it to stop'," *California Health Care Foundation*, May 24. https://www.chcf.org/blog/nurses-high-risk-suicide-just-wanted-all-stop/.

Gordon, James S. 2019. *The Transformation: Discovering Wholeness and Healing After Trauma*. San Francisco: HarperOne.

Greeson, Jeffrey M., Michael J. Toohey, and Michelle J. Pearce. 2015. "An adapted, Four-Week Mind-Body Skills Group for Medical Students: Reducing Stress, Increasing Mindfulness, and Enhancing Self-Care." *Explore* (NY) 11(3):186-192. https://doi.org/10.1016/j.explore.2015.02.003.

Gregg, Melanie. 2020. "The Breakthrough I've Been Waiting For: College Writing and Graphic Memoirs of Illness and Caregiving." *Graphic Medicine* 2022. https://vimeo.com/725873876/da6a636aa0.

Guillemin, Marilys and Lynn Gillam. 2015. "Emotions, Narratives, and Ethical Mindfulness." *Journal of the Association of American Medical Colleges* 90(6):726–31.

Haddad, Lisa M., Pavan Annamaraju, and Tammy J. Toney-Butler. "Nursing Shortage. [Updated 2022 Feb 22]." *StatPearls [Internet]*. Treasure Island, Florida: StatPearls Publishing. https://www.ncbi.nlm.nih.gov/books/NBK493175/.

Hass-Cohen, Noah, Joanna C. Findlay, Richard Carr, and Jessica Vanderlan. 2014. "'Check, Change What You Need To Change and/or Keep What You Want': An Art Therapy Neurobiological-Based Trauma Protocol." *Art Therapy*

31(2):69–78. http://dx.doi.org/10.1080/07421656.2014.903825.

Hills, Marcia and Jean Watson. 2011. *Creating a Caring Science Curriculum: An Emancipator Pedagogy for Nursing.* New York: Springer Publishing Company.

Horton-Deutsch, Sara and Gwen D. Sherwood. 2017. *Reflective Practice: Transforming Education and Improving Outcomes.* Indianapolis, IN: Sigma Theta Tau International.

John, Stephanie. 2019. "Imposter Syndrome: Why Some of Us Doubt Our Competence," *Nursing Times*, January 28. https://www.nursingtimes.net/roles/nurse-educators/imposter-syndrome-why-some-of-us-doubt-our-competence-28-01-2019/.

Kagan, Paula N. 2006. "Jo Ann Ashley 30 Years Later: Legacy for Practice." *Nursing Science Quarterly* 19(4):317–327. https://doi.org/10.1177/0894318406293121.

Kaimal, Girija, Kendra Ray,and Juan Muniz. 2016. "Reduction of Cortisol Levels and Participants' Responses Following Art Making." *Art Therapy* 33(2):74–80. https://doi.org/10.1080%2F07421656.2016.1166832.

Kelsey, Catherine and Hayes, Sally. 2015. "Frameworks and Models – Scaffolding or Strait Jackets? Problematising Reflective Practice." *Nurse Education in Practice* 15(6):393–396. https://doi.org/10.1016/j.nepr.2015.05.006 Permalink.

Kinman, Gail, Sandra Leggetter, Peter A. Leggat, and Derek R. Smith. 2016. "Emotional Labour and Wellbeing: What Protects Nurses?" *Healthcare* 4(4):1-12.

Knaak, Stephanie, Ed Mantler, and Andrew Szeto. 2017. "Mental Illness-Related Stigma in Healthcare." *Healthcare Management Forum* 30(2):111-116. https://doi.org/10.1177%2F0840470416679413.

LaPenna, Denise and Joseph D. Tariman. 2020. "Art Therapy: A Literature Review of Efficacy in Improving Psychosomatic Symptoms in Patients with Cancer." *Clinical Journal of Oncology Nursing* 24(2):123–26.

Lutz, Gabriele, Christian Scheffer, Friedrich Edelhaeuser, Diethard Tauschel, and Melanie Neumann. 2013. A Reflective Practice Intervention for Professional Development, Reduced Stress and Improved Patient Care—A Qualitative Developmental Evaluation. *Patient Education and Counseling* 92(3):337–345. https://doi.org/10.1016/j.pec.2013.03.020.

MacLaughlin, Brian W., Dan Wang, Anne-Michelle Noone, Nan Liu, Nancy Harazduk, Michael Lumpkin, Aviad Haramati, Pamela Saunders, MaryAnn Dutton, and Hakima Amri. 2011. "Stress Biomarkers in Medical Students Participating in a Mind Body Medicine Skills Program." *Evidence-based Complementary and Alternative Medicine* 2011(950461). https://doi.org/10.1093%2Fecam%2Fneq039.

Mehl-Madrona, Lewis. 2007. "The Nature of Narrative Medicine." *The Permanente Journal* 11(3):83-86. doi:10.7812/TPP/07-052

Mema, Briseida, Andrew Helmers, Cory Anderson, Kyung-Seo Min, and Laura E Navne. 2021. "Who am I? Narratives as a Window to Transformative Moments in Critical Care." P*loS One* 16(11):e0259976–e0259976. https://doi.org/10.1371/journal.pone.0259976 .

Monroe, Todd and Heidi Kenaga. 2011. "Don't Ask Don't Tell: Substance Abuse and Addiction among Nurses." *Journal of Clinical Nursing* 20(01).

Nolan, Gregory, Lucy Dockrell, and Suzanne Crowe. 2020. "Burnout in the Paediatric Intensive Care Unit." *Current Pediatrics Reports* 8(4):184–88.

Park, Kwi Hwa, Bee Sung Kam, So Jung Yune, Sang Yeoup Lee, and Sun Ju Im. 2022. "Changes in Self-Reflective Thinking Level in Writing and Educational Needs of Medical Students: A Longitudinal Study." *PloS One* 17(1):1-11. https://doi.org/10.1371/journal.pone.0262250.

Peng, Ying, Shao-Wen Xiao, Hui Tu, Xiao-Yun Xiong, Zhao-Jia Ma, Wen-Jun Xu, and Ting Cheng. 2022. "The Impostor Phenomenon among Nursing Students and Nurses: A Scoping Review." *Frontiers in Psychology* 13(809031): 1-8. https://doi.org/10.3389/fpsyg.2022.809031.

Saad, Lydia. 2022. "Military Brass, Judges Among Professions at New Image Lows." *Gallup*, January 12.

https://news.gallup.com/poll/388649/military-brass-judges-among-professions-new-image-lows.aspx .

Schmidt, Madeline and Kirstin Haglund. 2017. "Debrief in Emergency Departments to Improve Compassion Fatigue and Promote Resiliency." *Journal of Trauma Nursing* 24(5):317–322. https://doi.org/10.1097/JTN.0000000000000315.

Sierpina, Victor S., Mary Jo Kreitzer, Elizabeth Mackenzie, and Michelle Sierpina. 2007. "Regaining Our Humanity Through Story." *EXPLORE: The Journal of Science & Healing* 3(6):626–632. https://doi.org/10.1016/j.explore.2007.09.012.

Tervalon, Melanie and Jann Murray-García. 1998. "Cultural Humility Versus Cultural Competence: A Critical Distinction in Defining Physician Training Outcomes in Multicultural Education." *Journal of Health Care for the Poor and Underserved* 9(2):117–125.

U.S. Department of Health and Human Services, Office of the Surgeon General. 2022 "Addressing Health Worker Burnout; the US Surgeon General's Advisory on Building a Thriving Workforce" https://www.hhs.gov/surgeongeneral/priorities/health-worker-burnout/index.html.

van der Riet, Pamela, Tracy Levett-Jones, and Catherine Aquino-Russell. 2018. "The Effectiveness of Mindfulness Meditation for Nurses and Nursing Students: An Integrated Literature Review." *Nurse Education Today* 65:201–211. https://doi.org/10.1016/j.nedt.2018.03.018.

Weinlander, Eva E., Eric J. Gaza, and Marcy Winget. 2020. "Impact of Mind-Body Medicine Professional Skills Training on Healthcare Professional Burnout." *Global Advances in Health and Medicine* 9. https://doi.org/10.1177/2164956120906396.

Wexler, Teressa M., and Jana Schellinger. 2022. "Mindfulness-Based Stress Reduction for Nurses: An Integrative Review." *Journal of Holistic Nursing*. https://pubmed.ncbi.nlm.nih.gov/35213264/.

Yagelski, Robert. 2009. "A Thousand Writers Writing: Seeking Change through the Radical Practice of Writing as a Way of being." *English Education* 42(1): 6-28.

Yahyaei, Asma A, Alistair Hewison, Nikolaos Efstathiou, and Debbie Carrick-Sen. 2022. "Nurses' Intention to Stay in the Work Environment in Acute Healthcare: A Systematic Review." *Journal of Research in Nursing* 27(4):374-397. https://doi.org/10.1177/17449871221080731.

Zaharias, George. 2018. "What Is Narrative-Based Medicine?" *Canadian Family Physician* 64(3):176–180.

Zytkoskee, Adrian M. 2020. "Reflective Writing in Medical Education: A Writing Professional's Exploration of Purposes and Practices." Dissertation.

HJSR
50th Anniversary Edition

Cannabis, Communities, and Place: (Re)constructing Humboldt's Post-Prohibition Present

Joshua S. Meisel <meisel@humboldt.edu> (Cal Poly Humboldt), **Dominic Corva** (Cal Poly Humboldt), **Ara Pachmayer** (Cal Poly Humboldt)

Abstract

Since 1990, many Cal Poly Humboldt faculty and students have made cannabis the focus of scholarship and learning. This work has been shaped by the political, economic, and cultural legacies of cannabis in Humboldt County. Scholarly interest spans multiple dimensions of cannabis cultivation, commerce, consumption, and related social issues. As a multidisciplinary team of scholars, Cal Poly Humboldt faculty affiliated with the Humboldt Institute for Interdisciplinary Marijuana Research (HIIMR) have also shaped the Bachelor of Arts in Cannabis Studies that will launch in Fall 2023. This is the first social science degree program in the United States with this orientation. In this article we trace the origins of cannabis scholarship and learning at Cal Poly Humboldt from a period of domestic cannabis prohibition to the recent period of industry legalization. We also provide an overview of the BA program links to local and state policy and other contexts. Finally, we discuss cultivation communities and strategies of resilience in the context of economic turmoil as a result of legalization.

Keywords: Humboldt, cannabis, communities, place, post-prohibition

Over the past half century, Humboldt County has been significantly shaped by, and identified with, its relationship with cannabis cultivation. Statewide legalization presents new opportunities for the County and its communities to leverage this legacy and address challenges. Cal Poly Humboldt is located within this historical epicenter of cannabis cultivation. The campus is well-positioned to study cannabis from a range of disciplinary perspectives, as well as create a bachelor's degree for students to effectively address the challenges and opportunities that legalization presents.

In this article, we investigate entanglements of Humboldt County's transition from a state of prohibition to a post-prohibition present (Corva and Meisel 2022). As a team of scholars affiliated with the Humboldt Institute for Interdisciplinary Marijuana Research (HIIMR), we draw on our individual experiences and cumulative research experiences to consider the conjunctural crisis (see Hall and Massey 2010) of cannabis in the county.

In the next section below, we trace the history of cannabis scholarship and education at Cal Poly Humboldt. Later, we discuss statewide eradication campaigns and subsequent legalization that threatens the place of cannabis in the County plan for sustainable economic development. Next, we review the origins of cannabis cultivation in Humboldt County. We then describe research agendas examining cultivator organizing in the face of a market crash, as well as the transformation of cannabis communities by policy changes within and external to the County. In a subsequent section, we present a research agenda that compares Humboldt and Santa Barbara Counties' navigation of post-prohibition, particular policy contexts, and related implications for economic viability of Humboldt legacy farmers. Dillis et al. (2021) identify Santa Barbara as the geographic center of California "non-traditional" large-scale cultivation enterprises. Finally, we describe cultivation communities and their development of strategies of resilience to promote economic development.

Cal Poly Humboldt and Cannabis Studies Education and Research

In this section, we discuss the development of Cal Poly Humboldt cannabis education, as the state, county, and university grappled with vexing legal, economic, regulatory, public health, and environmental issues associated with legalization. Since 2002, the Department of Sociology at Cal Poly Humboldt has offered a weekend workshop, "The Emerald Triangle." Over the course of two days, students meet guest speakers representing cannabis farmers, public health agencies, city and county governments, land management agencies, and environmental organizations among other fields. Between 2011 and 2023, more than 700 students enrolled in the course. The curriculum provides an overview of social, economic, policy, and environmental dimensions of cannabis in the region. The course topics directly reflect emerging issues as the community navigated ambivalent relationships with the industry.

Between 2003 and 2019, 21 graduate students across academic programs completed cannabis-focused master's theses. Student questions about cannabis were also raised across classes and during office hours. In 2010 a core group of faculty from Anthropology, Biology, Economics, Environmental Science, the Library, Psychology, Social Work, and Sociology began meeting as a "Marijuana Studies Working Group." The group organized research presentations and attracted other faculty and students to participate. In 2012, the Chancellor approved their proposal to charter a research institute: "the Humboldt Institute for Interdisciplinary Marijuana Research (HIIMR)."

Josh Meisel (Sociology), our first author, and Erick Eschker (Economics) were the founding faculty co-directors until 2021 and 2019, respectively. Our second author, Dominic Corva (Cannabis Studies), is one of the current co-directors along with Whitney Ogle (School of Applied Health). Our third author, Ara Pachmayer (Recreation and Tourism), is a member of HIIMR. While the initial campus and CSU approval of HIIMR signaled important executive level recognition of the legitimacy of cannabis scholarship, substantial obstacles for research and teaching remained. The University retained prohibitions on research collaborations with the cannabis industry, as well as student internship and research placements with cannabis businesses. While there have been calls for eliminating such restrictions (Meisel, Watson, and Wesley 2019; Piomelli et al. 2019), these university prohibitions had a chilling effect on building the research capacity of HIIMR faculty and their students. In response, faculty sought to expand research partnerships within and external to the university. One way this was accomplished was by creating spaces where cannabis scholarship was shared with the broader campus community. In the first two years of HIIMR's existence, the co-directors organized 12 well-attended public lectures on a range of cannabis related topics.[1]

University support for cannabis scholarship and education declined between 2014 and 2019. During this period there was less active communication from university administration in support of research opportunities or executive level presence at HIIMR sponsored events. The tide changed again in 2019 when a new university administration reached out to Meisel and Corva, then co-directors of HIIMR, seeking HIIMR leadership on the development of an academic degree in Cannabis Studies.

1. A complete archive of these talks including video can be found here: https://humboldt-dspace.calstate.edu/handle/2148/1228

One year later the university was invited to submit a proposal to the Chancellor's Office of the California State University to become a polytechnic university. It is noteworthy that the invitation from the Chancellor (Cal Poly Humboldt 2020) referenced the need for academic programs to address a host of pressing needs in California, including "north coast crops," a veiled reference to cannabis cultivation. A new academic program in Cannabis Studies was included in the polytechnic proposal submitted in early 2021. The BA in Cannabis Studies was one of the first three new programs approved by the Chancellor's Office in 2022. Cannabis Studies was presented as foundational to this Cal Poly transformation.

HIIMR co-directors Meisel and Corva led the academic program proposal process with substantial input from HIIMR-affiliated faculty members. In the process, they outlined six substantive new core courses with cannabis-centered content anchoring two concentrations for undergraduates: *Environmental Stewardship* and *Equity and Social Justice*. Restrictions imposed by the California State University Chancellor's Office prevented curriculum from including any direct interaction with the plant or formal partnerships with any business that is directly involved in the cultivation, processing, transport, or distribution. Though these restrictions prevent teaching students some topics such as cultivation or manufacturing of products, faculty still had freedom to envision and design a curriculum unlike any of the trade-oriented programs that have emerged elsewhere nationally.

The BA in Cannabis Studies at Cal Poly Humboldt is an applied interdisciplinary critical studies program focused on developing graduates who can operate as historically informed stewards of cannabis legalization. From a critical perspective, cannabis legalization provides a material field for working on intersectional social problems through policy reforms in society, not simply a criminal justice reform in itself. Cannabis is a gateway subject for understanding and analyzing social problems. While legalization is a brand new tool for addressing such social issues, the broader context of prohibition remains an ongoing challenge.

As our students learn about the history, geography, culture, politics and botany of cannabis, they are prepared to steward legalization and other practices that mitigate cannabis-related problems, rather than reproduce them in new ways. Outside of Humboldt County, there are very few places in the world where this history is so evident in the very landscape, and woven into local activism and broader environmental social movements.

The study of cannabis is a vehicle for thinking about and practicing social change. The legacy of cannabis in Humboldt County makes it the ideal place to learn. To understand the origins of cannabis scholarship and learning at Cal Poly Humboldt requires an awareness of the history of cannabis prohibition—and legalization—in Humboldt County. This history highlights the ways in which the County was both insulated and exposed to the deindustrialization and economic decline common to rural landscapes across the world in the age of neoliberalization since the 1970s (Corva 2008).

As we continue to build the Cannabis Studies Program, we draw on a rich variety of methods for doing practical work on social problems. We depend on expertise across academic programs from hydrology and geospatial analysis to grant writing, investigative journalism, and beyond. Our students will have great flexibility in applying skills valued in many professions beyond those associated with cannabis policy, regulatory compliance, and socially conscious industry development. Ultimately, the study of Cannabis Studies at Cal Poly Humboldt prepares students to understand and address social problems broadly.

Cannabis and the Shaping of Humboldt County

The rise and fall of cannabis as an industry in Humboldt County in many ways mirrors the trajectory of other extractive commodity booms in the region (e.g., gold, whaling, fish, and timber) that sacrificed ecological sustainability and community stability for unsustainable economic growth fueling settler colonialism (Reed 2022). The industrialization of cannabis cultivation in the County, however, is not the primary reason why the county at large became culturally associated with cannabis cultivation, nor how the cultural economy of cannabis livelihoods worked for environmental and community sustainability rather than against it, for decades.

"Origin" stories centering cannabis and community formation in Southern Humboldt have been told many different times and ways throughout the decades. The story tellers were often the people who lived through it (Raphael 1985; Anderson 1990; Anders 1992; Holmquist 2010), thoughtful journalists like Emily Brady (2013), and more recently academics (Corva 2014; Kelly and Formosa 2022). Bosk (2000) provides a collection of personal interviews with back-to-landers that provides a glimpse of the communitarian values of the "new settlers" who first planted cannabis and/or became environmental activists in the county in the 1960s and 70s. A more complete account of this history can be gleaned from the 159 separate issues of "New Settler Interviews" published between 1985 and 2017. Raphael (1974, 1985, 1994, and 2022) provides extraordinary "everyday history" for linkages between the "old settler" colonial past of Humboldt County, starting with the Gold Rush, and the "new settlers" of the Bosk interviews.

In this article, we use the terms "new settler" and "back-to-the-lander" interchangeably and to emphasize three things. First, there is continuity of "settlerism" as a colonial process based on taking, owning, and exploiting land as private property (Reed 2022). Second, we make a cultural distinction between the "old settler" Humboldt families and the "new" ones establishing roots in the 1960s and 1970s. The latter generated terrains of cultural synthesis and contestation that continue to shape county communities today. This distinction is also between 1850s-style genocidal dispossession of the indigenous people and 1970s-style efforts to "inhabit a separate country" (Berg, ed. 1978) following principles of environmental stewardship. However, both old and new settlers have in common a more-than-proprietary love for the land and common struggle to stay here.

This section relies on those sources, as well as Corva's 14-year history as a field researcher in the County. The first period of this work, between 2009 and 2012, focused directly on the back-to-the-landers. Corva explored the relationships between 1960s social movements and "mom and pop" cannabis markets. By 2010 these markets were a minoritarian but vocal element of Humboldt's medical cannabis industry. The second period, between 2013 and 2016, found Corva in conversation with and then enlisted by elements of the Humboldt cannabis industry that organized county and state stakeholders in the lead-up to legalization. Corva served, for example, as the accompanying local academic expert for the Blue Ribbon Commission and the California State Board of Equalization tours of the Humboldt cannabis industry in 2014.

As noted, the story of the back to the land movement in Southern Humboldt includes how they accidentally found that cannabis could be a cash crop to sustain rural livelihoods and create community infrastructure. Cannabis producers provided in every remote watershed they settled both money and time to build enduring community institutions: schools, community centers, and volunteer fire departments. In the Garberville/Redway area that served as a peri urban center for most, they built a credit union, a rural health center, and the Mateel Community Center. The latter doubled as a center for social services, as well as an event and meeting venue for coordinating broad community development institutions and nonprofit fundraising events such as the *Reggae on the River* music festival. They formed enduring environmental nonprofits and wrote grant applications to professionalize them.

The new settlers did not build a singular community that identified itself as cannabis growers—certainly not in public. They built a network of institutions where community formation, defense, and resilience strategies could be worked out. And until 2010, they never organized under the banner of a singular cannabis industry. Even their community defense strategies against eradication revolved around avoiding arrests through early detection of incoming raid teams (the foundation of the local KMUD radio station's original programming); and protecting their civil rights as citizens and county rights as landowners.

This logic of community formation based on communitarian and environmental stewardship never went away. It was the foundation upon which the county became culturally tolerant of cannabis cultivation. It was a livelihood that was not necessarily connected to industrial-scale greed and environmental destruction. That said, federally-funded eradication inflated the street price of cannabis and created a financial incentive for people inside and outside of Humboldt County to cultivate cannabis for profit. Cannabis cultivation by a wide range of actors, including "old settler" families, new migrants escaping urban life, environmentally destructive opportunists, and criminal organizations domestic and foreign, spread throughout the county.

During the most intense period of federally funded eradication in Humboldt County (between 1985 and 1997, approximately), the development of cultivation techniques associated with industrial cannabis followed techniques for evading eradication. "Diesel dope" (cannabis grown indoors, off grid, and reliant on diesel powered generators) powered rural grows throughout the 1990s, as cultivators went indoors to escape helicopter detection. Cultivating on public lands mitigated the risk of planting on one's own property once asset forfeiture laws were passed in the mid-1980s. The strategy of growing on remote public lands was taken up by organized (and disorganized) crime. By the end of the 1990s, these grow locations tangentially fueled a racialized moral panic about Mexican cartel growers, a subject that is beyond the scope of this paper.

Increasingly large-scale cultivation pushed further and further into remote, ecologically sensitive watersheds where clear-cutting and industrial pesticide use helped maximize yields (Butsic and Brenner 2016; Butsic, Schwab, Baumann, and Brenner 2017). And the more cultivation sites that popped up, large-scale and otherwise, the more water was diverted to grow cannabis rather than sustain sensitive species habitats (Bauer et al. 2015). In the decade following the 1996 passage of the medical cannabis initiative, Proposition 215, enforcement spread out across the state and focused increasingly on large-scale, environmentally destructive grows on public lands (Corva 2014).

While Proposition 215 created new strategies for growers to protect their crops from eradication and, increasingly, decriminalized medical cannabis dispensaries on the West Coast, the industrialization of the cannabis cash crop exploded. Two pull factors besides Proposition 215 are of special note. First, in 2002 the county elected a progressive District Attorney, Paul Gallegos. He brought close ties to the cannabis-supported environmental nonprofits and activism in the area. In 2003, Gallegos issued the "99 plant" guideline which allowed each medical cannabis patient, or their caregiver, to grow up to 99 plants in a 100 square feet of canopy, without being prosecuted. While formally about indoor medical cannabis grows, as it was attached to 100 square feet of canopy, the guideline came to assume a sort of informal regulatory power throughout the County. And second, the financial crisis of 2007-2008 attracted economic refugees into jurisdictions that already tolerated medical cannabis activity.

All of this accelerated the trend of a commodity boom that, under legal circumstances, would have played out in years rather than decades, into a supply glut. More and more people came to grow cannabis in Humboldt, and fewer people were punished for growing. The evolution of cannabis from a hippie cash crop to an industrial commodity boom also meant that it became a significant part of the county economy, valuable as a bulwark against rural poverty and stagnation. But because the value of cannabis derived from the price support program of prohibition, the boom times lasted for decades.

In 2010, with cannabis legalization on the horizon, fear of "what's after pot" became a subject of popular anxiety for cultivation communities in Southern Humboldt (Brady 2013). At the same time, the boom was entering its zenith: margins fell but the volume and velocity of money it generated commanded an ever-increasing, if hard to measure, share of the county economy. Conservative estimates put the share at 25% of county GDP (Budwig 2013). The paradox of the moment is that it pushed cultivators, for the first time, to gain formal recognition from the County in order to create a sustainable, regulated economic future for an industry whose value was greatly inflated by prohibition. Unsustainable revenue and production figures were being used to project the size and significance of the industry, and to predict tax revenue to be collected from the industry; as well as to propose regulations that limited the viability and profitability of the legal industry.

The period between 2010 and 2016 accelerated structural tendencies towards extractive, large-scale cultivation, even as many supported the legalization of cannabis in order to reign it in. While small-scale, longtime communitarian and environmentalist grower communities struggled to reconcile economic livelihoods with the industrial character of what had grown around them, a new wave of commercially oriented market actors arrived in the county in advance of and/or to gain a foothold in legal cannabis production.

The 2014-2016 period saw unprecedented organization of cultivators as an *industry* (as opposed to civil defense groups, rural property owners, environmental activists, small business owners, and nonprofit communitarians funded by a cash crop). The new organization prepared the county for its present position: a place that remains central to the state's cannabis system, as measured by a number of comparative measures to be discussed in the next section. The effort to represent cannabis as a modern, industrial partner in the county's regulation of cannabis was catalyzed by an organization called California Cannabis Voice Humboldt (CCVH), created by two Bay Area lawyers and funded largely by commercial cannabis growers in the county,[2] and at times in tension with small scale livelihood growers associated with the values and remaining people of the back to the land movement.

CCVH initiated the mobilization of industry capital into political capital for Humboldt County legalization, in the California legislature, for the state's Board of Equalization, and for then-Lt. Governor Gavin Newsom's Blue Ribbon Commission (BRC). The BRC informed the 2016 medical cannabis legislation, both the Medical Marijuana Regulation and Safety Act and the Medical Cannabis Regulation and Safety Act. CCVH convened an open forum of cultivators and county stakeholders to create the first draft of Humboldt County's first ever cannabis regulation ordinance. The Humboldt County Board of Supervisors revised the policy after considerable pressure emerged from environmentalist and small-scale cultivation voices concerned about the scale of industry to be permitted, as well as its lack of environmental review.

The first county ordinance regulating cannabis cultivation, referred to by county officials as "Ordinance 1.0," drew upon local input, and subsequently that local input informed California's guidelines for structuring regulations. Of particular note is the BRC report's declared commitment to transitioning "responsible actors"; protecting public lands and sensitive watersheds; and protecting small and medium sized entities (SMEs) from large corporate interests (Newsom, Soltani, and Humphreys 2015). The last point is especially salient, since the 2017 Medicinal and Adult Use Cannabis Regulation and Safety Act launched with a five-year ban on large license types.

The initial ordinance created a window for transitioning existing cultivation sites by December 31, 2016, with a mandate to conduct an Environmental Impact Report (EIR) that could support a more comprehensive land use ordinance (termed Ordinance 2.0) to accommodate new permit applications. More than 2,200 applications were received, most of which failed to advance to state licensure within a few years, and the county went to work on an EIR to support a second, less rushed cultivation ordinance that could establish land use regulations for new cultivation sites on ecologically appropriate parcels (Ford 2020).

Based on the EIR, as well as the location of pre-2016 cultivation sites governed by Ordinance 1.0, the Planning Department capped by watershed the number of cultivation permits and canopy acreage.[3] Ordinance 2.0 also established a temporary moratorium on permits in areas of Yurok Traditional Tribal Cultural Affiliation. Overall, 3,500 permits and 1,205 acres were distributed across the county. More than a third of the permits were planned for the Cape Mendocino and South Fork Eel watersheds where Southern Humboldt's earliest cultivation communities settled.

Ordinance 2.0 also extended the period for 1.0 permits to come into compliance with state and local regulations. A 2020 "small cultivator" amendment established a streamlined permitting process for existing cultivation under 2,000 square feet on homesteads. Broadly speaking, the County's approach to the state's

2. This account of CCVH—what it was, who drove it, where the fault lines were—is abbreviated for space. The story is much more complex and will be worth a separate historiographic intervention.

3. Canopy acreage refers to the cultivation square footage permitted by license type.

personal use cultivation and medical cultivation is more generous than anywhere else in the state. Beyond the licensed landscape, Humboldt maintains a relatively permissive approach to cultivation that allows its citizens to grow their own or as caregivers for patients with doctor's recommendations.

Like every other jurisdiction in California, the permitting process has been slow and difficult to navigate. But by state comparison, Humboldt was much more prepared for and effective at getting permits through the process. It was aided by revenues from the county's cultivation tax, Measure S, an initiative created by its cultivation communities and passed in 2016. Measure S funds allowed the Planning Department to double its budget and hire 50 new employees (Ford 2020).

Measure S also funded the county's aggressive abatement program by allowing the Sheriff's Department to hire two more deputies and paying for a satellite-based Humboldt Environmental Impact Reduction (HEIR) enforcement program, for which it received an award from the California State Association of Counties. This aggressive code enforcement program combined with punitive fines focused disproportionately on Southern Humboldt, relative to the rest of the county, with ambivalent results (Norris 2021). In 2020, it was lauded by the County Sheriff and the local licensed cannabis trade association, the Humboldt County Grower's Alliance (HCGA), for substantially reducing unregulated cannabis cultivation in the county, and reducing negative community impacts and the county homicide rate (Greenson 2020).

The HEIR program also, however, created substantial collateral damage to people, mostly located in Southern Humboldt's oldest cannabis cultivation watersheds, who were *not* large-scale, environmentally destructive, or even current cannabis cultivators, creating a public outcry in Southern Humboldt. A series of articles by investigative journalist Nichole Norris drew the attention of a national libertarian nonprofit, the Institute for Justice, which commenced a lawsuit against the county in Fall 2022 (Norris 2022).

The promise of cultivation taxes as a reliable source of county tax revenue were first diminished by another Southern Humboldt-based lawsuit completed in 2021 (Norris 2021), then by the increasing failure of licensed cultivators to pay their cultivation taxes on time due to lower than projected revenues, and finally through two actions by the County Board of Supervisors. In February of 2022 the Board retroactively reduced 2020 Measure S rates by 85% and then in November suspended the tax completely for 2021-2022 (Hutson 2022).

At the time of this writing, Humboldt County faced a radically different set of conditions for its cannabis cultivation communities than it did in 2010, when those communities first contemplated the challenge of surviving legalization. There developed a productive tension between legacy small-scale cultivators associated with environmental stewardship, and large-scale cultivators materially capable of stewarding policy formation as an industry. The tension materialized in a licensed landscape characterized by diversity of ownership and a predominance of cottage and specialty businesses reflecting more closely the original scale of mom-and-pop growers.

Humboldt and Santa Barbara Counties: Case Studies of Dispersed versus Narrow Production Control

This section describes Humboldt's licensed cannabis landscape in relation to the state and by comparison with Santa Barbara County, the largest licensed cultivation jurisdiction in California. The data for this analysis were from the Department of Cannabis Control's publicly available "Search Tool" database[4] on October 30, 2022. There were 18 different state cultivation license types available through January 21, 2023. At that time "large" license types opened for application. This comparison focuses on the subset of cultivation license types available until January 2023: small, medium, specialty, and specialty cottage licenses.

While the 2017 state regulations capped at one the individual holding of medium license types, a loophole

4. https://search.cannabis.ca.gov/, accessed December 5, 2022.

Table 1. Comparison of Humboldt and Santa Barbara Licensed Cultivation

	Humboldt County		Santa Barbara County		California
	N	% of State Total	N	% of State Total	N
Cultivation Canopy Acres	413	15.3	437	16.2	2,693
Business Owners	861	21.0	72	1.8	4,108
Total Licenses	1,485	13.9	1,875	17.5	10,712

allowed an individual to own unlimited non-medium license types (Dillis et. al. 2021). As a result, cultivation firms could "stack" smaller license types to produce far beyond the individually owned maximum canopy size intended by Proposition 64 that legalized recreational cannabis. This regulatory glitch bypassed California's five-year moratorium on large license types defined as more than one acre in size. The ban on holding more than one medium license type remained intact.

While managing similar acreage, Humboldt County cannabis production control is dispersed across many more business owners compared to the narrowly controlled Santa Barbara market (Table 1). Many cultivation firms, including those in Humboldt, chose to stack small license types. One fifth of state producers are located in Humboldt with approximately 1.7 licenses per business owner. In contrast, Santa Barbara producers hold on average 26 licenses per business owner. As a result, Santa Barbara's 1,875 licenses are held by 72 unique business owners (1.8% of the state total) permitted to grow up to a total of 437 acres of canopy. In comparison, Humboldt's 1,485 licenses are held by 861 unique business owners (21.0% of the state total) permitted to grow up to a total of 413 acres of canopy.

About 50% of Santa Barbara County licenses (n=980) are owned by five individuals, compared with about 10% (n=83) of Humboldt licenses owned by the county's top five license holders. Santa Barbara has one dominant license type, the stackable "small" one. Almost all (98%) of Santa Barbara's cultivation licenses are small outdoor (full sun) (n=1,409) and Small Mixed Light Tier 1 (small low energy light deprivation) (n=425) cultivation types (Table 2). And the top five license holders in Santa Barbara have both more licenses and more cultivation canopy area than Humboldt's largest license holders. Humboldt has a range of producer sizes within the county, but those producers are not very large grows in the state context.

There is a greater diversity of license types in Humboldt County (see Table 2) and more unique owners than in Santa Barbara, although the maximum canopy size at the time of this writing is almost equal (Table 1). Together Humboldt and Santa Barbara account for more than 30% of California's maximum potential licensed canopy acreage. License ownership in Santa Barbara is consolidated amongst far fewer owners, reflecting a degree of license stacking into a few industrial mega-grows that reflect corporate and investor ownership rather than small- and medium-sized enterprises. Humboldt's significance to state totals comes from its many producers with disproportionately less licensed canopy cultivated per owner. There is approximately a half-acre per owner in Humboldt compared to 6.1 acres in Santa Barbara.

The concentration of ownership and diversity of license type is what distinguishes Humboldt from Santa Barbara County. About 60 percent of California's Medium Mixed Light Tier 1 licenses (low energy use light deprivation) are in Humboldt. Additionally, 1 in 4 of the state's largest single license type, one-acre medium outdoor licenses (full sun), and 19% of the state's medium sized, high energy, licenses (Specialty Cottage Mixed Light Tier 2), are located in Humboldt County. Humboldt also has a significant share of medium license types, which are the largest scale cultivation licenses that can be owned by any one business in California. It should be noted that all of these are outdoor medium license types.

Table 2. Top 10 Cultivation License Types Permitted in Humboldt and Santa Barbara Counties (Ordered from highest to lowest cultivation canopy permitted)

	Humboldt County		Santa Barbara County		California (Total)
License Type	N	% of State Total	N	% of State Total	N
Medium Outdoor	102	24.9	7	1.7	409
Med Mixed Light Tier 1	142	58.9	5	2.1	241
Small Outdoor	272	5.8	1,409	30.3	4,651
Small Mixed Light Tier 1	517	20.6	425	16.9	2,509
Specialty Outdoor	108	24.8	6	1.4	435
Specialty Mixed Light Tier 1	159	40.3	1	0.3	395
Specialty Mixed Light Tier 2	21	31.8	0	0.0	66
Specialty Mixed Light Tier 2	21	31.8	0	0.0	66
Specialty Cottage Mixed Light Tier 1	68	35.6	1	0.5	191
Specialty Cottage Mixed Light Tier 2	5	19.2	0	0.0	26
Med Mixed Light Tier 2	13	18.8	2	2.9	69

But Humboldt also has a disproportionate share of California's extra-small (specialty) and extra-extra small (specialty cottage) license types. These license types represent the legacy of the early cannabis farmers insofar as they reflect smaller scale operations. In other words, Humboldt's presence in legal cannabis can be understood in terms of both the number of owners, as well as the area of licensed canopy. Humboldt County growers hold about 25% of California specialty outdoor licenses, about 40% of specialty light deprivation licenses, and 32% of specialty high energy outdoor licenses (Table 2). Similarly, the county hosts about 35.6% of California specialty cottage light deprivation licenses and about 19% of specialty cottage high energy outdoor licenses. Its 12.7% share of specialty cottage outdoor licenses is in line with its overall license share.

The state license database as of October 30, 2022 reflected a snapshot of a regulated landscape that is both the outcome of a process that began as early as 2010, and a starting point for preserving the dream of legal cannabis as a significant economic industry in Humboldt County. That significance is financial, but it is also cultural. As an outcome, it reflects a local public policy trajectory aimed at constructing an industry commercial enough to matter for local economic development (mediums and stacked smalls), but culturally distinct enough to capitalize on its legacy of livelihood growers (unstacked smalls and all smaller license types) that made it all possible in the first place.

Cannabis and Humboldt County Sustainable Economic Development

Sustainable economic development calls for economic growth to work in tandem with sustaining and enhancing society and the natural environment for future generations. Economic development that embraces sustainability cannot ignore people and place. According to the U.S. Economic Development Administration, sustainable economic development should include the unique assets and strengths of a region (U.S. Economic Development Corporation, n.d.). In the case of Humboldt County, this includes cannabis.

Telling the stories of the people and places of Humboldt County is increasingly an important consideration for economic development in the county. Every five years the Humboldt County Economic Development Division, GoHumCo, develops a Comprehensive Economic Development Strategy (CEDS) "to attract federal funding and technical assistance" (Anon 2023). The most recent CEDS process focused on storytelling related to the economic drivers of Humboldt County and solicited feedback from communities and the public. This process resulted in a report that will inform the final CEDS document (Anon 2023). As an activity linked to the federal level where cannabis remains illicit, the final CEDS cannot reference cannabis explicitly. This federal constraint is problematic for planning as it limits discourse on cannabis and economic development. In the current document cannabis is referred to instead as "alternative agriculture" (Humboldt County Economic Development 2018:6).

Federal prohibition poses an objective challenge for cannabis industries in Humboldt County and in the United States in general. Until federal prohibition ends, or interstate trade agreements are enacted, a national legal cannabis market allowing for export beyond California is not possible. In the interim, opportunities exist for integrating the cannabis economy into Humboldt County. Prospective directions for cannabis and sustainable economic development discussed below include normalizing cannabis tourism, reducing the stigma of cannabis, and strategic partnerships. Potential community and environmental impacts are also discussed.

While Colorado became the first state to offer recreational sales in 2014, legal cannabis tourism is a relatively new activity in the United States. Since 2014, multiple states have legalized recreational sales, and cannabis tourism has evolved. Taylor defined cannabis tourism "as purchasing with the intent to consume marijuana products while temporarily traveling away from one's normal place of work or residence" (2019:6). Giraudo (2019) goes beyond consumption to include those interested in learning about cannabis but perhaps not consumption. Both ways of considering cannabis tourism provide opportunities for Humboldt County

Despite being the epicenter of cannabis cultivation, cannabis tourism was slow to develop in Humboldt County largely due to regulations limiting farm visits or complex permits for tour providers. Currently, cannabis tourism in Humboldt County resembles an agritourism experience based on guided tours with visits to farms and/or nurseries to learn about cannabis and cultivation. Tours end in a visit to a dispensary to purchase cannabis, often from the farmers visited on the tour. There are multi-day all-inclusive tours, cannabis-themed events, or cannabis options at events being offered more frequently as regulations allow. Cannabis may also be combined with existing tourism resources in Humboldt County such as redwood forests, beaches, and other natural areas, as tourists find cannabis-friendly accommodations, and experience dispensaries, cannabis lounges, or events independently.

In addition to the potential of agritourism for Humboldt cannabis, showcasing best practices in cannabis cultivation for tourists, potential investors, and industry professionals is a possible area of growth and provides a model for others to see and learn about cannabis culture in the county. Huckleberry Hill Farms touts a Fish Friendly Farm designation and Full Cup Farm highlights their regenerative practices. By linking best practices in cannabis cultivation with best practices at traditional farms, visitors are provided a comprehensive perspective of sustainable agriculture in Humboldt County that includes cannabis.

Legal cannabis tourism has the potential to spread development to other businesses as well. In 2021 there was an estimated $25 billion in legal cannabis sales in the United States. According to Forbes "as much as $4.5 billion was driven by tourists, who pour an additional $12.6 billion into restaurants, hotels, attractions and other shops—as well as into state and municipality tax coffers" (Yakowicz and Kelleher 2022). More locally, in a 2018 report on Humboldt County's New Cannabis Landscape, tourism was noted as an important industry for cannabis (Claesgens and Kraft 2018). Claesgens and Kraft found that cannabis tourism would not only impact the demand for Humboldt cannabis but also "… for Humboldt-branded products and services" (2018:9), which would ultimately lead to job creation in the tourism sector.

Along with federal prohibition, overcoming the stigma of cannabis is a challenge for sustainable economic growth, whether the stigma originates with other residents (Kang and Lee 2018; Pachmayer, Switzer, and Reilly 2021), the government, businesses, or other tourists. Efforts to reduce the stigma of cannabis may be formal, such as New York state's educational media campaign launched in advance of legal sales, or the Visit Modesto Cannapass program that focused on making "sure the community and users are all comfortable" (Kudialis 2021). Other efforts to reduce stigma may be more informal, such as in the context of tourism. And still other efforts elevate products and places through boutique branding and spa-like environments for consumption.

Looking to research from Canada where cannabis is federally legal, Dupej and Nepal (2021) investigated the role of tourism in normalizing cannabis. The authors found that by incorporating cannabis as another tourism resource, the tourism industry is facilitating the normalization of cannabis. "Tourism places cannabis into full public view, giving it greater visibility as a legitimate recreational resource" (Dupej and Nepal:367). Cannabis lounges introduce ways to consume in a safe environment, while cannabis tours to dispensaries offer a structure for people to learn how to purchase cannabis. Dupej and Nepal specifically address the messages provided by cannabis tour guides which challenge the deviant image of cannabis. Reviews of Humboldt Cannabis Tours highlight the educational factor of the tour and multiple reviews mention the tour guide role in dispelling myths about cannabis and Humboldt County. In the long term, focusing on destigmatizing cannabis coupled with fostering a legal path federally may allow the cannabis industry to grow. There will likely be continued prohibition at the federal level without education and work towards destigmatizing cannabis.

Strategic partnerships are important for a host of reasons, including the sharing of resources, knowledge transfer and innovation, as well as access to new markets or funding opportunities (Rainey et al. 2003). Strategic partnerships may also strengthen the cannabis industry. For example, there exist partnerships between cannabis and non-cannabis businesses for marketing purposes. These partnerships help integrate cannabis businesses into the traditional business world, providing exposure to non-profit organizations and educational institutions for creative funding partnerships, research, sponsorship, and advocacy opportunities. As noted in a marketing assessment, "collaboration between industry and government should be the foundation of Humboldt's cannabis marketing program" (Gordon and Carver 2020:69). Based on their research of regionally branded products such as Colombian coffee and Napa Valley wine, the authors noted that industry and government collaboration was necessary for a marketing program that would successfully promote a regional product and promote the region itself. Industry stakeholders have knowledge and expertise while governments determine policy and address industry related issues.

While sustainable economic development may result in an increase in job opportunities that can improve the standard of living of residents, there are other potential impacts on a community including environmental protection, enhanced community participation and solidarity, and improvement in the quality of life. As established above, there is a strong connection between the cannabis communities of Humboldt County and the natural environment that encourages environmental stewardship. There are a variety of environmental impacts often attributed to illegal cannabis grows including water diversion violations, illegal road construction, discharge of sediment and pollutants to waterways or watersheds, or threats to native species (Bauer et al. 2015; Butsic and Brenner 2016). As the Department of Cannabis Control for the state of California operates with strict regulations regarding environmental impacts of cultivation, legal farms face regulations to ensure the negative impacts of their activity are reduced, ultimately protecting our shared natural resources.

As noted above, community participation is a critical part of the CEDS process. In soliciting feedback, GoHumCo notes:

> We intentionally began this process by listening. To create an effective and inclusive economic development strategy, we need to know what community members are experiencing, their priorities, and their insights on the strengths and weaknesses of our county (We are Humboldt Rising, n.d.).

Rainey et al. (2003) considered the importance of equity in community participation and recommended that people "across all segments of the community" must be involved in the development process (2003:713). This sense of empowerment in promoting the success of a community enhances feelings of community solidarity and investment as evidenced by the lengthy history of community supporting organizations throughout the county. Strong community links also facilitate the formation of the types of partnerships noted above (Weinberg 2000).

At this time, the cannabis industry in Humboldt County is imperiled (Vanderheiden 2023). In better years, the tax revenue generated by the industry had the potential to improve the quality of life of all residents. Revenue flowed into community services, investments in infrastructure, healthcare, and other public goods (Rainey et al. 2003), as well as targeted investments and improvements in the cannabis industry. While the benefits of tax revenues from the cannabis industry are shared throughout the county, the cannabis industry is most directly responsible for and impacted by the work of many of the actions noted in this section.

Humboldt County at the Crossroads

The Humboldt County cannabis economy is at a crossroads. This represents the latest chapter in the political, cultural, and economic evolution of Humboldt County growers. Coupled with the high cost of entry into regulated markets—where those markets actually exist—there are multiple forces further constraining licensed and unlicensed cannabis as a significant component of the Humboldt County economy. The annual eradication program that began in 1983, California's Campaign Against Marijuana Planting (CAMP) has been reinvigorated and rebranded as the Eradication and Prevention of Illicit Cannabis (EPIC) program. EPIC expanded from a seasonal to a continuous cannabis statewide interdiction program targeting unlicensed growers (Thompson 2022).

Additionally, a group of county residents successfully placed an initiative on the 2024 ballot to suspend cultivation licensing in the county (Burns 2022). The anger catalyzing initiative organizing mirrored that of neighborhood frustration with Arcata indoor cultivation of the prior decade. Collectively these developments signal a new chapter in the demonization of cannabis farmers in Humboldt County. These events suggest that the cannabis sector continues to face exceptional challenges to gaining legitimacy.

Whether cannabis remains an economically viable Humboldt County agricultural commodity may hinge on cultivator capacity to influence the cannabis policy dialogue in the years ahead. The market crisis has weakened the financial ability of cultivators to act politically based on strictly financial claims about market power and promises of significant tax revenue. Unlike other jurisdictions like Santa Barbara with concentrated market control, life in Humboldt is highly linked to broad cannabis ownership and a particular environmental and community building ethic from which the local industry evolved. This Humboldt cannabis story continues to shape critiques of the negative impacts of cannabis broad commercialization during later iterations of the Green Rush.

Every corner of Humboldt County has been touched by the long and mostly informally regulated cannabis boom. While the new settlers began growing for cultural reasons, the material windfall of cannabis as a cash crop carried the plant through an entire war on drugs and on to the other side of a formally regulated industry. Humboldt's insurgent cannabis communities "overgrew" the government. They held back the counterinsurgent forces of drug crop eradication administered globally by the United States. In the process, political strategy positioned marginalized communities as public enemies of a war for public health (Corva 2008). This narrative of cannabis cultivation as an anti-drug war social movement captures some but not all that made cannabis central to a rural capitalism industrial commodity boom.

Humboldt County's challenge today is to rescue the past communitarian values in service of charting a path forward for sustainable, inclusive, economic development. Yet the cannabis industry cannot stand alone as the engine of economic development. Ancillary value chains beyond crops must be nurtured to create a sustainable future. This work also requires a diverse, sustainable, economic sector related to the legacy of cannabis in the county.

This sustainable pathway requires creating ways for people and small businesses to generate meaningful livelihoods out of other cannabis licenses and ancillary services: manufacturing, tourism, events, retail, breeding, product development, marketing and the like. It means supporting the capacity of farms to differentiate their products from those found in a mass commodity market. Medium and small farms cannot compete with large industrial-scaled farms. They have to provide a product of materially greater quality (a craft industry like micro-brewed beer) with niche cultivars. And they have to provide a product that can use collective marketing strategies to differentiate their products (like environmentally-friendly certification and wine appellations), with the knowledge that their customer base is a niche rather than mass market.

While many can and are doing this on their own or in specialized groups, the county can help by supporting, for example, education in regenerative agricultural methods such as those featured in second-generation Southern Humboldt filmmaker Clair Weissbluth's documentary "Tending the Garden" (2022). Regenerative methods are marketable and reduce input costs over the long term, as they rely on locally developed inputs rather than mass-produced fertilizer products. This is a "craft beats scale" approach. Another resilience strategy underway within the county is happening through a nonprofit called the Cooperative Agriculture Network (Cory 2022), which involves forming cooperative arrangements to scale elements of business plans independent of production, such as distribution and marketing. The county's local equity program has been supporting this effort by funding licensees to receive training in cooperative business strategies.

While the county collective marketing program seems to be defunded at this time, the Humboldt County Grower's Alliance is proceeding on their own, based on the marketing assessment they wrote with a county contract (Gordon and Carver 2020). That assessment, in turn, was informed by collaborative expertise from The Organization for an International Geographical Indications Network (oriGIn), an NGO based in Geneva, Switzerland; and Origins Council (OC), a 501(c4) association of legacy producing jurisdictions in California. HCGA is now a member of OC, and as such helped found a national advocacy organization called the National Craft Cannabis Coalition. The group lobbies, among other things, for a national system of indicators of origin when federal policy changes. Independent of the county, its main trade association is strategizing to create appellations marketing frameworks that have served other industries, most famously wine, quite well. Eventually, when California figures out how to implement regional appellation certification, it will build on efforts currently underway in Humboldt and its OC partners.

Indirectly, the county can support its cultivation communities by finding industry-specific and broad-based ways for people and small businesses to be resilient in their livelihood strategies. For cultivation communities, this especially means policy flexibility to fallow licenses without losing them permanently, and opportunities to develop complementary income streams from other crops or even other jobs, perhaps in the restoration sector. Of course, Humboldt's communities beyond cannabis cultivation are also experiencing economic hardship. Broad-based workforce development plans that can help individuals, families, and small businesses in the county become resilient, regardless of their participation in cannabis, are absolutely necessary.

Ultimately, the value of cannabis to Humboldt County is not simply the economic value of its cultivation. Economically, the task at hand is to institutionalize the socially positive legacies of cannabis in the county into a heritage—a renewable cultural resource for differentiating Humboldt as a place that showed the world glimpses of how destructive, dominant social orders could be challenged and changed. As such, it is the place where the limits of legalization as an end to the drug war are especially evident. Humboldt's historical cannabis legacy is not over: it is just beginning to be written as a series of lessons to learn about how to be resilient in the face of destructive forces seemingly beyond our control—for now.

Cal Poly Humboldt: Research and Education in the Eye of the Storm

The launch of the Cannabis Studies major at Cal Poly Humboldt coincides with the moment when the statewide implementation of legalizing cannabis is facing strong headwinds from its enduring entanglements with prohibition. These include the retail bottleneck across the state; the expensive and complex regulatory environment; the broken promises of protecting small and medium sized enterprises; the emergence of not-in-my-backyard politics even in Humboldt County; and ongoing federal prohibition preventing legal access to national markets and limited access to banking. It is more apparent than ever that we need to study the ontology of legalization rather than assume it means a clear break from prohibition. Cal Poly Humboldt faculty and students remain physically and intellectually in a hub for studying and informing these changes. Despite the impacts of an industrialized prohibition economy, the legacy of communitarianism and environmental stewardship create unique threads in the Humboldt story of place and people. Our new polytechnic institution is growing a multidisciplinary intellectual community with strong ties to policy, practice, and change. These Cal Poly Humboldt faculty and students are well positioned to inform the national conversation that will shape the political, cultural, and economic landscapes of cannabis post prohibition.

References

Anders, Jentri. 1990. *Beyond Counterculture: the Community of Mateel.* Pullman, WA: Washington State University Press.

Anderson, Mary Siler. 1990. *Whatever Happened to the Hippies?* San Pedro, CA: R&E Miles.

Anon. n.d. "We Are Humboldt Rising." Retrieved January 22, 2023. https://www.humboldtrising.com/.

Anon. 2022. "Marijuana Enforcement Team Finds 'Numerous' Environmental Violations Involving Nearby Watersheds While Raiding Illegal So-Hum Grows, HCSO Says." *Lost Coast Outpost* June 8. Retrieved March 20, 2023. https://lostcoastoutpost.com/2022/jun/8/marijuana-enforcement-team-operation-in-southern-h/.

Bauer Scott, Jennifer Olson, Adam Cockrill, Michael van Hattem, Linda Miller, Margaret Tauzer, and Gordon Leppig. 2015. "Impacts of Surface Water Diversions for Marijuana Cultivation on Aquatic Habitat in Four Northwestern California Watersheds." *PLoS One* 10(9):e0120016.

Berg, Peter, ed. 1978. *Reinhabiting a Separate Country: A Bioregional Anthology of Northern California.* San Francisco: Planet Drum Foundation.

Bosk, Beth Robinson, ed. 2000. *The New Settler Interviews Vol. I: Boogie at the Brink.* White River Junction, Vermont: Chelsea Green Publishing Company.

Brady, Emily. 2013. *Humboldt: Life on America's Marijuana Frontier.* First Edition. Boston: Grand Central Publishing.

Budwig, Jennifer. 2013. "Potential Economic Impact to Humboldt County if Marijuana is Legalized."

Presented at the Humboldt Institute for Interdisciplinary Marijuana Research Speaker Series, Arcata, CA, February 26. https://scholarworks.calstate.edu/concern/publications/h128nh33t.

Burns, Ryan. 2022. "Supes Agree to Put Controversial Weed Initiative on 2024 Ballot, Though They Hope to Work with Organizers on Alternatives." *Lost Coast Outpost.* Retrieved October 26, 2022. https://lostcoastoutpost.com/2022/oct/26/supes-agree-put-controversial-weed-initiative-2024/.

Butsic, Van and Jacob C. Brenner. 2016. "Cannabis (Cannabis sativa or C. indica) Agriculture and the Environment: A Systematic, Apatially-Explicit Survey and Potential Impacts." *Environmental Research Letters* 11:044023.

Butsic, Van, Benjamin Schwab, Mathias Baumann M, and Jacob C. Brenner. 2017. "Inside the Emerald Triangle: Modeling the Placement and Size of Cannabis Production in Humboldt County, CA USA." *Ecological Economics* 142:70-80.

Cal Poly Humboldt. 2020. "Humboldt State Asked to Plan for Consideration to Be State's Third Polytechnic." November 23. Retrieved December 4, 2022. https://now.humboldt.edu/news/humboldt-state-asked-to-plan-for-consideration-to-be-states-third-polytechn.

Claesgen, D. and M. Kraft. 2018. *rep.* Eureka, CA: Humboldt County Workforce Development Board.

Corva, Dominic. 2008. "Neoliberalization and the War on Drugs: Transnationalizing Illiberal Governance in the Americas." *Political Geography* 27(2):176-193.

Corva, Dominic. 2014. "Requiem for a CAMP: The Life and Death of a Domestic U.S. Drug War Institution." *International Journal of Drug Policy* 25(1):71–80. https://doi.org/10.1016/j.drugpo.2013.02.003.

Corva, Dominic and Joshua S. Meisel, eds. 2022. *The Routledge Handbook of Post-Prohibition Cannabis Research.* New York: Routledge.

Cory, Oliver. 2022. "CAN Ready to Take Clients, Trellis Funding Now Available: Cooperation Humboldt." *Redheaded Blackbelt.* September 22 Retrieved March 29, 2023. https://kymkemp.com/2022/09/12/can-ready-to-take-clients-trellis-funding-now-available-cooperation-humboldt/.

Dillis, Christopher, Eric Biber, Hekia Bodwitch, Van Butsic, Jennifer Carah, Phoebe Parker-Shames, Michael Polson, and Theodore Grantham. 2021. "Shifting Geographies of Legal Cannabis Production in California." *Land Use Policy* 105:105369. doi: 10.1016/j.landusepol.2021.105369.

Dupej, Susan and Sanjay K. Nepal. 2021. "Tourism as an Agent of Cannabis Normalization: Perspectives from Canada." *Tourism Review International* 25(4):353–69.

Giraudo, Rachel F. 2019. "Blazing a Trail: Cannabis Tourism in the United States." Pp. 105–12 in *The Politics of Marijuana: A New Paradigm*, edited by T. McGettigan. Peter Lang Inc. International Academic Publishers.

Gordon, Ross and Terra Carver. 2020. rep. Eureka, CA: Humboldt County Economic Development Department. Retrieved June 2, 2022. https://www.gohumco.com/DocumentCenter/View/226/Humboldt-County-Cannabis-Marketing-Assessment.

Greenson, Thaddeus. 2020. "'Crawling in the Right Direction' A Look at the State of the Local Cannabis Industry from Bookending Perspectives." March 18. Retrieved March 28, 2023. https://www.northcoastjournal.com/humboldt/crawling-in-the-right-direction/Content?oid=16848658.

Hall, Stuart and Doreen Massey. 2010. "Interpreting the Crisis: Doreen Massey and Stuart Hall Discuss Ways of Understanding the Current Crisis." *Soundings* (44):57-68.

Holmquist, Scott. 2010. *Chronic Freedom.* Arcata, California: Open Secret.

House, Freeman. 1999. *Totem Salmon: Life Lessons from Another Species.* Boston: Beacon Press.

Humboldt County Economic Development. 2018. "Prosperity! 2018: Comprehensive Economic Development Strategy 2018-2023." Humboldt County, California. https://www.gohumco.com/DocumentCenter/

View/137/2018-to-2023-Comprehensive-Development-Strategy-PDF.

Hutson, Ryan. 2022. "Humboldt County Supervisors Amend Measure S, Remove Cultivation Tax for 2022 and 2023." *Redheaded Blackbelt*. Retrieved January 31, 2023. https://kymkemp.com/2022/11/05/humboldt-county-supervisors-amend-measure-s-remove-cultivation-tax-for-2022-and-2023/.

Kang, Soo K. and Jaeseok Lee. 2018. "Support of Marijuana Tourism in Colorado: A Residents' Perspective Using Social Exchange Theory." *Journal of Destination Marketing & Management* 9:310–19.

Kelly, Erin Clover and Marissa Lia Formosa. 2022. "Two Rural Industries Intersecting Over Time: Cannabis Production and Ecological Restoration in the Mattole Valley, California, USA." Pp. 266-274 in *The Routledge Handbook of Post-Prohibition Cannabis Research*, edited by Dominic Corva and Joshua S. Meisel. New York: Routledge.

Kudialis, Chris. 2021. "City In California Launches First U.S.-Based Marijuana Tourism Campaign." Leaf Report August 1. Retrieved June 1, 2022. https://www.leafreport.com/news/city-in-california-launches-first-u-s-based-marijuana-tourism-campaign-12183.

Meisel, Joshua, Tara Marie Watson, and Jared Wesley. 2019. "Eliminate Restrictions Preventing Cannabis Research." *Policy Options*. November 22. Retrieved March 28, 2023. https://policyoptions.irpp.org/magazines/november-2019/eliminate-restrictions-preventing-cannabis-research/.

Newsom, Gavin, Abul Soltani, and Keith Humphreys (Steering Committee). 2015. *California Blue Ribbon Commission Pathways Report*. Safeandsmartpolicy.org. Retrieved March 20, 2023.https://newrevenue.org/2023/03/30/california-blue-ribbon-commission-report-on-marijuana-legalization/.

Norris, Nichole. 2021. "Permitted Cannabis Farmers Win Case Versus Humboldt County Yesterday." *Redheaded Blackbelt*. Retrieved January 31, 2023. https://kymkemp.com/2021/03/12/permitted-cannabis-farmers-win-case-versus-humboldt-county-yesterday/.

Norris, Nichole. 2022. "Humboldt County Cannabis Abatement Program 'Unconstitutional," says the Institute for Justice Which Today Filed a Federal Class Action Lawsuit." *Redheaded Blackbelt*. Retrieved January 31, 2023. https://kymkemp.com/2022/10/05/humboldt-county-cannabis-abatement-program-unconstitutional-says-the-institute-for-justice-which-today-filed-a-federal-class-action-lawsuit/.

Pachmayer, Ara, Reed Switzer, and Shawn Reilly. 2021. "Cannabis Tourism and the Community: Resident Attitudes in Humboldt County, California." *IdeaFest: Interdisciplinary Journal of Creative Works and Research from Humboldt State University* 5(1).

Piomelli, Daniele, Robert Solomon, Donald Abrams, Agnes Balla, Igor Grant, Thomas Marcotte, and John Yoder. 2019. "Regulatory Barriers to Research on Cannabis and Cannabinoids: A Proposed Path Forward." *Cannabis and Cannabinoid Research* 4(1):21–32. doi: 10.1089/can.2019.0010.

Rainey, Daniel, Ralph Christy, Allen Ivye, and Kenneth Robinson. 2003. "Essential Forms of Capital for Sustainable Community Development." *American Journal of Agricultural Economics* 85(3):708–15.

Raphael, Ray. 1974. *An Everyday History of Somewhere*. New York: Knopf.

Raphael, Ray. 1985. *Cash Crop: An American Dream*. Mendocino, CA, Redway, CA: Ridge Times Press.

Raphael, Ray. 1994. *More Tree Talk: The People and Politics of Timber*. Island Press.

Raphael, Ray. 2022. *"I Like it Here": Life Stories of Humboldt's Bob McKee*. Arcata: Creative Type Productions.

Reed, Kaitlin. 2022. "Cannabis, Settler Colonialism, and Tribal Sovereignty in California." Pp. 53-62 in T*he Routledge Handbook of Post-Prohibition Cannabis Research*, edited by Dominic Corva and Joshua S. Meisel. New York: Routledge.

Taylor, Lorraine. L. 2019. "Defining Marijuana Tourism." *Journal of Hospitality & Tourism Research* 43(3):438–46.

U.S. Economic Development Corporation. n.d. "About Overview." Retrieved April 13, 2023. https://www.eda.gov/about.

Vanderheiden, Isabella. 2023. "'We're in Crisis Mode': Collapsing Cannabis Industry Guts Garberville Businesses; Local Leaders Look to Tourism as a Saving Grace." *Lost Coast Outpost.* January 2 Retrieved March 20, 2023. https://lostcoastoutpost.com/2023/jan/2/garberville-photos/.

Weinberg, Adam. 2000. "Sustainable Economic Development in Rural America." *The Annals of the American Academy of Political and Social Science* 570(1):173–85.

Yakowicz, Will and Suzanne Rowan Kelleher. 2022. "Cannabis Tourism Is Now A $17 Billion Industry—And It's Just Taking Off." *Forbes* May 29. Retrieved June 3, 2022. https://www.forbes.com/sites/willyakowicz/2022/05/29/cannabis-tourism-is-now-a-17-billion-industry-and-its-just-taking-off/?sh=58659cda2056.

HJSR

50TH ANNIVERSARY EDITION

Remembering Sam Oliner: Legacy in the Memories and Lives of his Students and Friends

We want to honor Dr. Samuel P. Oliner, one of the founders of HJSR and a Professor of Sociology at Humboldt. Dr. Oliner passed away November 17, 2021.

*The following text is from the nomination of Dr. Oliner in 2008 for the Humboldt County Peacemaker Prize.**

After the extermination of his family and community, Samuel P. Oliner survived the Holocaust in Europe during World War II with the help of a Polish farm family. This experience guided his development and passion for understanding how good can triumph over evil. His early research was on rescuers of Jews during the Holocaust, leading to later work on racism, ethnicism such as anti-Semitism, prejudice, and discrimination. His recent work has been on altruism, caring, heroism, apology, and forgiveness and on the promotion of behaviors congruent with these concepts. His most recent book, *Altruism, Intergroup Apology, Forgiveness and Reconciliation* (with Piotr Zylicz), is currently in press. Now retired as a professor from Humboldt State University, he is an internationally known scholar, researcher, and teacher whose list of publications and presentations fills six single-spaced pages and includes authorship of numerous books, chapters, articles, and book reviews. His life and work have inspired productions and documentaries in the U.S. and abroad including, for example, the BBC documentary *To Give or Not To Give*. He is often interviewed by domestic and international journalists for radio and television. His ethic of caring and compassion has permeated his life's work as a role model, mentor, and teacher.

Sam taught at Oakland Community College before coming to HSU in 1971. He is an Emeritus Professor of Sociology at HSU and often is invited to give lectures and workshops. His areas of expertise include such topics as the Holocaust, collective behavior, social movements, race and ethnic relations, intergroup apology and forgiveness, altruism and prosocial behavior, minorities in Eastern Europe and the former USSR, heroism, and social change. He and his wife Pearl Oliner were honored as Scholars of the Year at HSU after the publication of their seminal work T*he Altruistic Personality: Rescuers of Jews in Nazi Europe*. He founded the *Humboldt Journal of Social Relations* in 1973 and the Altruistic Personality and Prosocial Behavior Institute in 1982 at HSU. His editorship of the *Journal* has enabled it to survive as a forum for encouraging the publication of social science research. Since retiring from HSU, he has continued his work as director of the Institute which is an important vehicle for teaching research skills to graduate students and exploring important research topics. Sam has successfully applied for many research grants over the years enabling the Institute to employ undergraduate and graduate students. He continues to foster and promote students' research into altruism, prosocial behavior, apology, and forgiveness, including many research projects in the local community.

Sam is a valuable resource to our community, his former students, and colleagues. He has served as a significant and effective role model for his students and other audiences who have been inspired by his presentations. Many teachers have been greatly influenced by his mentoring. His former students are teaching at College of the Redwoods, HSU, and public schools in Humboldt County as well as throughout the state and country. Stephen Stamnes is a former student and mentee of Sam's and is a lecturer at HSU who teaches a class entitled *Sociology of Altruism and Compassion*, a course that was originally created and taught for many years by Sam. Steve noted, "My students in the altruism class are profoundly moved by Sam's contributions to the promotion of altruism and prosocial behaviors. Inspired by Sam's model of living a life infused with an ethic of care and compassion, an impressive number of them choose to enter the caring professions of teaching, nursing, and social work, and go on to serve this community."

Sam generously gives his time speaking in local elementary schools, high schools, and colleges, promoting the message of creating a more just and inclusive society. His book, *Toward A Caring Society: Ideas into Action*, written with his wife Pearl Oliner, noted that one way of encouraging the creation of a more peaceful and just society is by reaching out to young people. As he and Pearl wrote in *Toward A Caring Society*, "Study and teaching cannot be entirely the same if words like 'love' and 'community' become as well accepted as 'academic excellence,' 'merit,' and 'competence.'" Sam lives those values whether he is talking and working with children at Blue Lake School or providing on-going support and mentoring to graduate students during his retirement. He is constantly invited to give lectures on subjects related to how we might build a more caring, inclusive, and peaceful community. He has volunteered to give numerous guest lectures and workshops at HSU on such topics as apology and forgiveness, altruism and caring, and the Holocaust and genocide, to name a few. He has also presented at College of the Redwoods on such subjects as genocide, racism and genocide, altruism, and apology and forgiveness. According to Linda Ellis, his former student who currently teaches at CR, "Students at CR have been inspired by his presentations on apology and forgiveness. Sam is often unaware of the positive impact that his work on apology and forgiveness has had on students with whom he speaks. They have shared with me how inspired they are with his message and that it could possibly be a way to mend a broken world." He has spoken to elementary and high school students in Arcata, Eureka, McKinleyville, and Fortuna, engaging them in discussions about good and

Celebrating Oliner 2008 Peacemaker Prize (from left to right) Jason Whitley (MA 2009), Steve Stamnes (MA 1993), Dr. Jack Schaffer, Dr. Sam Oliner, & Dr. Josh Meisel

evil; altruism; improvement of social relations by treating others with respect and tolerance; the attributes of caring people; and deconstruction of evil, prejudice, and destructive behaviors.

Sam has also benefited the broader community. His current work on apology and forgiveness has direct applicability to inter-group relations between European American ethnic groups and the various indigenous Tribal groups of Humboldt County. He has spoken recently to the Arcata Open Door Clinic staff on altruism and caring, as well as at the Unitarian Church and the Westhaven Center for the Arts on apology and forgiveness. He conducted a workshop on altruism and forgiveness for the Osher Lifelong Learning Institute. He participated in a recent Channel 13 program on love and altruism and has volunteered to present at conferences on racism, ethnicity, and cultural inclusiveness at HSU. He is a Board member of Temple Beth El where he teaches Yiddish classes, and he is involved in raising funds in our local community to assist people who have been victims of the genocide in Darfur.

Because of Sam's abundant contributions to our community in promoting caring, apology, and forgiveness, and in providing us—children and adults alike—with a lens through which to understand how human goodness can prevail, he is an impressive and worthy candidate for the 2008 Humboldt County Peacemaker Prize.

**This nomination letter is reprinted here with permission of authors Stephen Stamnes, MA 1993 and retired faculty member in Sociology, Cal Poly Humboldt, and Ann Diver-Stamnes, Professor Emeritus of Education, Cal Poly Humboldt. Many students and faculty were co-signers of this nomination letter. Sam Oliner was awarded the Peacemaker Prize on November 19, 2008.*

Times - Standard 12/6/08 B-1

Professor promotes peace

Sam Oliner honored for his work dealing with altruism and reconciliation

ARCATA — Humboldt State University Professor Emeritus Sam Oliner, an internationally recognized author, peace ambassador and founder of the Altruistic Personality & Prosocial Institute, has been awarded the 2008 Humboldt County Peacemaker Prize by the Eureka Interfaith Fellowship.

The $4,000 cash prize is awarded biennially and honors individuals who have made substantial contributions to fostering peace in the Redwood Coast region.

Sam Oliner

A survivor of the Holocaust and the author most recently of "Altruism, Intergroup Apology, Forgiveness and Reconciliation," Oliner joined Humboldt State as a professor of sociology in 1971. He and his wife, Pearl, are former HSU Scholars of the Year and he founded the Humboldt Journal of Social Relations in 1973.

His former students teach at Humboldt State, College of the Redwoods and public schools across Humboldt County, the state and the nation, carrying his message of altruism and reconciliation.

Oliner continues to speak regularly at area schools and colleges and previously taught a workshop on altruism and forgiveness for Humboldt State's Osher Lifelong Learning Institute, which serves adults over the age of 50.

A board member of Temple Beth El, where he teaches Yiddish, Oliner helps raise funds locally to assist the victims of the ongoing genocide in Darfur.

Oliner 2008 Peacemaker Prize - Eureka Times-Standard newspaer clipping entitled "Professor promotes peace." Dated December 6, 2008.

Remembering Sam Oliner:
Legacy in the Memories and
Lives of His Students

It's hard to think of the words "prosocial" and "heroes" without an instant smile and remembering Sam—God gets to enjoy Sam's wisdom now! Thank you Sam!

— Gregory C. Gibson, Ph.D.
Professor of Sociology, Kent State University
MA in Sociology (2002), Cla Poly Humboldt

I think I speak for many when I say that Sam was not only a beacon of hope in the field of altruism and forgiveness, but he was a mentor, friend, and like family to so many. He is dearly missed and never forgotten.

— Mark Bauermeister, Ph.D.
Instructor, University of San Francisco
and Foothill College
MA in Sociology (2007), Cal Poly Humboldt

It is difficult to find words that adequately capture who Sam Oliner was as a person, and his scholarship. I worked with Sam at the Institute in Founders Hall from 2009-2011. He started out as my boss, but quickly became my mentor and my family. Sam had a hard time saying "no" to the many schools, teachers, and reporters that would ask him to retell his life story. It was because he believed it was important that people knew about such atrocities, so society would not repeat the same hate. He also spent an equal amount of time reminding people that in the midst of hate, the world is still filled with goodness and beauty. Sam chose to live in the light and shine that light on those around him. He was a lot of things to a lot of people—a scholar, Professor, a Holocaust survivor, a mentor, and a solid human being.

Sam gave me a lot of lessons about research, scholarship, and life. There are way too many to write here, but one of the most important lessons he ever gave me was, "Never shake hands like a guppy." For two years, I would walk into the office, he would stand up, and we would shake hands. Thanks to Sam, I have a solid handshake.

— Amanda Bertana, Ph.D..
Assistant Professor of Sociology,
Southern Connecticut State University
BA and MA in Sociology (2007, 2011),
Cal Poly Humboldt

Sam Oliner was my mentor and friend. I had the good fortune of first working with Sam during my graduate work. He was teaching a course on the Holocaust and graciously agreed to be on my thesis committee. I later became a co-managing editor of the HJSR with my good friend, Mark Bauermeister. As young graduate students, myself, Mark and another colleague, Amanda Bertana, worked with Sam in the Altruistic Personality & Prosocial Behavior Institute, recently renamed the Altruistic Behavior Institute.

As a descendant of an Armenian Genocide survivor, I was captivated by Sam's own story of surviving the Holocaust and his ability to teach such difficult topics with consistency, conviction, and compassion. My graduate work entailed combatting denial of the Armenian Genocide. When I faced threats from genocide deniers, it was Sam who told me, "You cannot quit. They want you to quit and if you do, they win. Do not let them win." I can chart my path of weaving together activism and academia directly back to that moment in Sam's office. He encouraged me to continue writing so that I could use my platform to educate and agitate.

Sam championed his graduate students and continued to mentor us and collaborate long after our graduation and his retirement. When I began teaching sociology, Sam was my first guest lecturer. His book *Narrow Escapes: A Boy's Holocaust Memories and Their Legacy* (2000) would be the inspiration for my own course, "Narrating Genocide." Sam and Pearl Oliner were prolific writers and the scholarship they produced, especially in the field of Altruism and Compassion, is unparalleled. My own scholarship is heavily influenced by Sam, which is why I mention him in every invited lecture I give.

I owe a great deal to Sam, professionally and personally, which is why I am so proud to be working alongside my colleague, Dr. Ronnie Swartz, Director of the Altruistic Behavior Institute. We are keeping the conversations on altruism going and in making sure that Sam and Pearl's contributions remain accessible to folks around the world. Sam was an exceptional scholar and teacher. I loved him and my life has forever been changed because of him. May Sam and Pearl's memory be a blessing. And may the *Humboldt Journal of Social Relations* and the Altruistic Behavior Institute continue to educate, inspire, and create community for another 50 years!

— **Maral N. Attallah, M.A.**
Distinguished Lecturer,
Department of Critical Race,
Gender & Sexuality Studies,
Cal Poly Humboldt
BA and MA in Sociology (2005, 2007),
Cal Poly Humboldt

Sam Oliner in Lower Library Office with beloved Greek sailor's cap (circa 2000)

HJSR

50TH ANNIVERSARY EDITION

Cal Poly Humboldt

Sponsored Programs Foundation

LOCAL RESEARCH. GLOBAL REACH.

WHO ARE WE?

Sponsored Programs Foundation (SPF) is a non-profit auxiliary housed within the Cal Poly Humboldt Office of Research. SPF supports grant-funded research and programs on campus by providing grant and contract services. SPF administers virtually all externally-funded grants/contracts and submits proposals to funding agencies on behalf of Cal Poly Humboldt.

RESOURCES TO GET STARTED

SPF is available to assist you at each step along the way! Learn more about us, our services, and different resources for finding funding by visiting **our website!**

SUCCESS SNAPSHOT

622 Number of active projects SPF currently manages

140M Current total value of all active projects

61M Dollar amount of awards received in FY 21/22

CONNECT WITH US!